The Dystopian Mind

From Fire to AI: The Pattern Behind Every Human Revolution

by Shrikar Nag

Copyright © 2026 by Shrikar Nag

Published by Critical Angle Publications

ISBN 979-8-9956651-0-6 (paperback) ISBN 979-8-9956651-1-3 (ebook)

Published in the United States of America

First Edition

10 9 8 7 6 5 4 3 2 1

For Arya and Arha.

You were born into an age when many of the strangest systems had already become ordinary.

This book is for you because the future that will shape your lives may not arrive as something frightening at first. It may arrive as something convenient, intelligent, useful, beautiful, and difficult to refuse. My hope is that you grow up able to see clearly when the world is changing beneath your feet, especially when that change no longer looks strange to anyone else.

May you always keep the rare human instinct to notice when the normal is new.

What shocks one generation becomes infrastructure for the next.

What enters as convenience can leave as destiny.

Table of Contents

Preface

This book began with a pattern hiding in plain sight.

Again and again, across wildly different eras, human beings encountered changes that first looked unnatural, unsettling, or morally difficult to absorb. Then those same changes became useful. Then normal. Then invisible. The more I looked across history, the harder it became to ignore the repetition.

This book is an attempt to trace that repetition with seriousness and clarity.

It is written for readers who care about history, technology, psychology, power, and the future, but who also care about form. The argument here is ambitious by design. The subject moves from prehistory to artificial intelligence and possible singularity. To make that span meaningful rather than loose, the book is organized around one governing question: what human baseline did each revolution rewrite, and how did that rewritten baseline become ordinary enough to stop feeling chosen?

This book is written as trade nonfiction rather than academic history. That means it aims to be rigorous without becoming dry, literary without becoming vague, and provocative without sacrificing evidence. Each chapter is deepened with documented sources, notes, counterarguments, and chapter-level factual scaffolding.

The claim of this book is not that all progress is false, nor that every technological change is dystopian. The claim is narrower and more unsettling: civilizations repeatedly normalize transformations faster than they fully understand them, and the deepest consequences of those transformations are often psychological before they are political.

The book is therefore meant to be read with two sensations at once: recognition and unease. Recognition, because much of what it describes is already part of ordinary life. Unease, because ordinary life may be changing in ways that are easiest to see only after they have become difficult to refuse.

That claim matters now because the revolutions underway do not feel distant. They feel intimate. Systems that influence thought, work, intimacy, identity, and decision-making are no longer futuristic abstractions. They are part of ordinary life. That is precisely why they require historical perspective. The more intimate a transformation becomes, the easier it is to mistake it for reality itself rather than recognize it as a condition that could once have felt strange.

If this book succeeds, it will leave the reader with a different way of seeing both history and the present. It will make familiar developments feel structurally legible. It will make convenience feel morally charged again. And it will sharpen the reader's ability to ask, in every era, the same difficult question: what are we becoming willing to call normal?

The best outcome for a book like this is not agreement alone. It is residue. A sentence remembered in the middle of an ordinary day. A device that suddenly feels less innocent. A convenience that becomes harder to consume without thinking about its price.

Introduction

The Future Does Not Arrive When It Is Invented. It Arrives When It Stops Feeling Strange.

In the summer of 2023, a woman in a mid-sized American city sat at her kitchen table and asked a machine to write a eulogy for her father.

She did not know how the machine worked. She did not know what a large language model was, or how neural networks processed language. She knew only that she was grieving, that she could not find the words, and that someone had told her this thing could help.

She typed a few sentences about her father, his love of gardening, his quiet humor, his habit of reading the newspaper aloud at breakfast. The machine returned three paragraphs of polished, emotionally resonant prose. It captured something real. Not because it understood her father. It had never met him. But it had processed enough human language to produce patterns that felt like understanding.

She read it twice. She cried. She used most of it at the funeral. Afterward, she felt grateful and slightly uneasy, though she could not have said exactly why.

That unease is what this book is about.

Not the machine. Not the technology. Not even the specific capabilities of artificial intelligence. What this book is about is the thing that happened next — the thing that did not happen. The woman did not stop using the tool. She did not feel the need to tell anyone at the funeral that a machine had helped write the words. Within weeks, she was using it for work emails, for her son's college application essays, for a complaint letter to her landlord. The strangeness faded. The utility remained. The extraordinary became administrative.

That sequence — disturbance, utility, absorption, invisibility — is the oldest pattern in human civilization.

There is a way to test this claim immediately, before reading another page. Put this book down for a moment and try to remember what life felt like before the smartphone.

Not the facts, you know the facts. You know that before 2007 there was no iPhone, that before 2010 smartphones were not ubiquitous, that before 2012 social media had not yet saturated daily experience. You know the timeline. But can you remember the feeling? Can you reconstruct, from the inside, what it felt like to sit in a waiting room with nothing to do? What it felt like to be lost in an unfamiliar neighborhood without GPS? What it felt like to not know the answer to a question and to simply not know it, to let the not-knowing sit there, unresolved, until you happened to encounter the answer or forgot the question entirely?

If you are old enough to remember the pre-smartphone world, you may find that the feeling is surprisingly difficult to reconstruct. Not because your memory is poor, but because the current state of things, permanent connectivity, instant access, continuous partial attention — has become so thoroughly the default that imagining its absence requires genuine cognitive effort. You know, intellectually, that you once lived differently. But the embodied memory of that difference has faded. The previous baseline has been overwritten.

That overwriting is what this book is about.

If you are too young to remember the pre-smartphone world, the demonstration is even more striking. For you, permanent connectivity is not a state you adopted. It is the world. The idea that people once navigated by paper maps, waited days for letters, and experienced idle moments without reaching for a device is not threatening or disturbing. It is simply quaint, the way older generations talk about walking uphill in the snow. The strangeness has been completely absorbed. The revolution is complete.

And yet the revolution happened within living memory. The world that smartphones created is less than two decades old. In the long sweep of the history this book traces — from fire to AI, from the first campfire to the possible singularity — the smartphone era is a blink. But the speed with which its strangeness was absorbed is itself a data point. It tells us something about the velocity at which the pattern can operate in the modern world. It tells us that the interval between "this is strange" and "this is normal" is shrinking with each successive revolution. And it tells us that the next revolution, whatever form it takes — may be absorbed even faster.

That acceleration is one reason this book was written. Not because the present is uniquely dangerous, but because the present is uniquely fast. The time available for conscious evaluation — for seeing the costs alongside the benefits, for noticing what is being traded, for choosing rather than merely adapting, grows shorter with each revolution. If there was ever a moment when awareness mattered, it is now. Not because the future can be stopped. But because the future might still, with enough clarity, be chosen.

The most powerful revolutions in human history are rarely the ones people recognize in time.

They do not always arrive as catastrophe. They do not always look like conquest. More often, they enter as convenience, relief, efficiency, or prestige. They solve a problem that feels urgent, remove a friction that feels unnecessary, or offer an advantage too visible to ignore. At first, they may seem unnatural, excessive, invasive, or morally difficult to absorb. But if they provide enough utility, enough seduction, or enough competitive force, the same sequence begins to repeat.

First, they disturb us.

Then they help us.

Then they disappear into the structure of life.

That is how the deepest transformations win.

Human beings like to tell a cleaner story about civilization. We call it progress. We describe history as a long ascent from darkness into knowledge, from fragility into control, from scarcity into power. In one sense, that story is true. We learned to make fire, cultivate land, build cities, write laws, mint currency, print ideas, harness electricity, mechanize labor, digitize communication, and create machines that now appear to reason, generate, summarize, predict, and decide. These revolutions expanded possibility. They lengthened life, increased scale, accelerated trade, multiplied knowledge, widened power, and redefined what human societies could coordinate.

But they did something else as well.

Every revolution carried a hidden lesson. Every leap forward taught human beings not only what they could do, but what they could grow used to. Fire did not merely light the night; it altered the relationship between human beings and natural limits. Agriculture did not simply produce food; it reorganized time, labor, settlement, inheritance, hierarchy, and dependence. Writing did not only preserve memory; it made command durable and debt difficult to escape. Industry did not just increase output; it trained the body to obey the clock. Smartphones did not merely connect people; they made human attention continuously available to extraction. Artificial intelligence may not simply automate tasks; it may redraw the psychological boundary around what humans still experience as uniquely their own.

This book is about that deeper process.

Not only invention, but normalization.

Not only transformation, but accommodation.

Not only what changed in the world, but what changed in the human mind so that the world's new conditions could be endured, accepted, and eventually mistaken for reality itself.

Civilization advances not only through breakthroughs. It advances through repeated acts of psychological adjustment. Again and again, human beings encounter a force that unsettles their habits, moral instincts, or inherited idea of the natural. They resist it for a time. Then they absorb it under pressure, temptation, necessity, ambition, imitation, or relief. Then they reorganize around it. Then they forget they ever lived differently.

What begins as disruption becomes infrastructure.

What begins as artifice becomes environment.

What begins as a tool becomes a condition of existence.

The future does not conquer us when it is invented. It conquers us when it begins to feel normal.

This is one of the central patterns of human history, and one of the least honestly examined. We have many histories of invention, many histories of empire, many histories of ideas. We have far fewer histories of what human beings learn to tolerate, then accept, then defend. Yet this is often where the deepest civilizational change occurs. A thing that once felt invasive begins to feel practical. Then beneficial. Then necessary. Then obvious. By the time it becomes ordinary, it no longer appears as a choice at all. It appears as reality.

And once something feels like reality, its costs stop presenting themselves as costs.

Normalization is how change becomes invisible.

It is also how power becomes intimate.

The effect is why the word *dystopian*\\ in this book does not refer only to visible tyranny, cinematic collapse, or theatrical fantasies of machines ruling over human beings from some separate realm. The more durable versions are quieter. They arrive as habits, defaults, incentives, interfaces, conveniences, and systems that become too useful to reject and too familiar to see clearly. They are not always experienced as oppression. Very often they are experienced as improvement.

The effect is what makes them dangerous.

The greatest threat in a transforming civilization is not always that people will be forced into conditions they hate. It is that they will be gently guided into conditions they learn to love before they learn to question them.

The darkest institutions in history are not always the ones that terrify first. Often they are the ones that arrive too seductively to be named as danger while they are still entering the room.

Human beings do not drift into altered realities because they have a natural appetite for domination. They drift because each new arrangement offers something too compelling to dismiss: safety, speed, comfort, abundance, order, status, stimulation, relief, competitive advantage, or emotional ease. Every revolution removes a friction. But not every friction was merely an obstacle. Some frictions were forms of protection.

Distance once limited communication, but it also protected people from constant interruption. Slowness once frustrated action, but it also protected reflection. Scarcity of publication once constrained expression, but it also slowed the spread of contagious nonsense. Physical presence was once required for many forms of trust, intimacy, and proof. Human cognitive limitation was once an unavoidable fact, but it also preserved the experience of being mentally central to one's own world. If one wished to persuade, comfort, govern, or deceive another person, one generally needed more time, more labor, more proximity, more skill, or more visible effort than many systems require now.

When a revolution removes one of these frictions, it often produces unusual gains. It can also remove a restraint, a delay, or a difficulty that had been quietly preserving some portion of human sovereignty.

The point is why the story of progress is never only a story of liberation. It is also a story of substitution.

One burden disappears; another takes its place.

One limit falls; one dependence enters.

One discomfort is relieved; one form of autonomy narrows.

This does not mean revolutions are mistakes. It means revolutions are bargains, and civilizations are often poor negotiators when the immediate reward is vivid and the long-term price is diffuse.

We can see this pattern everywhere once we know how to look. Consider how swiftly the smartphone moved from marvel to extension of the self. Within a remarkably short span, a device that would once have appeared to earlier generations as a magical convergence of map, camera, marketplace, newspaper, social theater, work portal, private archive, and surveillance apparatus became something closer to an invisible baseline. Its absence now feels more unnatural to many people than its presence. That is normalization.

Consider artificial intelligence. For many people, the first encounter carries a jolt. A machine drafts coherent prose, produces code, summarizes law, generates images, suggests strategy, or imitates a human style with startling fluency. The first response is often wonder or unease. Yet once the system becomes useful enough, the emotional arc changes. The disturbance softens. The workflow adapts. The remarkable becomes administrative. What once felt uncanny becomes a tab in a browser, a button in a product, an assistant in a meeting, a baseline expectation in an industry. That is normalization.

Now consider older revolutions. The first printed books did not feel natural. The idea that text could be copied at scale altered religious authority, public argument, and access to belief itself. Industrial time did not feel natural either. Human bodies had to be retrained around bells, clocks, shifts, and output. Monetary abstraction was no small change. It required people to accept that unlike things could be compared on a shared scale and that value could travel stripped of much of its original social context. Agriculture itself was a rupture. It replaced movement with settlement, variety with cultivation, flexibility with planning, and immediate ecological responsiveness with repetitive obligation.

Again and again, the same structure appears.

A condition once experienced as strange becomes functional.

Then ordinary.

Then indispensable.

Then forgotten as a condition at all.

This final stage is especially important. The last phase of normalization is amnesia. Once a institution becomes ordinary enough, people stop remembering the prior world clearly enough to judge what was gained and what was lost. They no longer feel that they are living inside an arrangement earlier humans might have found disorienting, dependent, or intolerable. They feel only that they are living inside reality. This is how entire societies become enclosed by their own adaptations.

A child born into the age of screens does not experience the smartphone as an invasion. It appears as background. A worker born into digital workflow does not experience permanent reachability as a historic anomaly. It appears as professionalism. A society born into platforms and predictive systems does not experience algorithmic mediation as a philosophical threshold. It appears as convenience.

That is how transformations disappear into ordinary life.

And that disappearance is itself one of history's most powerful political events.

A institution no longer needs to justify itself once it stops being felt as a system. It becomes the waterline of civilization.

One reason the present moment matters so much. Artificial intelligence, synthetic media, algorithmic prediction, behavioral surveillance, and digital dependence all feel unprecedented because they are unprecedented in form. But they are not unprecedented in structure. They belong to a much older story: the story of how human beings repeatedly acclimate to systems that alter freedom, labor, trust, intimacy, identity, and thought itself. The question is not whether this has happened before. The question is whether this time the pace of adaptation will outrun the pace of recognition.

A civilization can survive many revolutions. The harder question is whether it can remain conscious during them.

This book does not argue for nostalgia. It does not argue that humanity should have rejected cities, abandoned science, refused computation, or remained obedient to every natural limit. It is not a plea for primitivism, nor a simple denunciation of technology. The aim is more demanding than that. The aim is to make the present legible.

Because if previous revolutions changed what humans could do, the revolutions now underway may change what humans are willing to be.

The chapters that follow move through history chronologically, but their deeper movement is psychological. Each one examines a revolution not only as an event, but as a rewrite of a human baseline. Fire rewrites night. Language rewrites experience. Agriculture rewrites time. Cities rewrite belonging. Writing rewrites memory. Money rewrites worth. Industry rewrites rhythm. Media rewrite attention. The internet rewrites separation. Smartphones rewrite solitude. Artificial intelligence may be rewriting mental exclusivity, authorship, authenticity, and eventually the very boundary of the human itself.

Seen from that angle, history stops looking like a museum of inventions and starts looking like a sequence of internal edits.

What shocks one generation becomes infrastructure for the next.

What once required justification eventually requires no explanation at all.

What once looked like interference becomes expectation.

The present moment is therefore not exceptional in pattern, only in speed and in proximity to the species itself. Earlier revolutions changed environment, settlement, trade, communication, and work. The coming ones reach deeper. They approach cognition, companionship, authorship, perception, prediction, and personhood. The arc is no longer only outward. It is increasingly inward.

That is why the book begins where it does.

The story starts at the earliest threshold: the moment human beings first discovered that the world did not have to be accepted on its own terms. Before cities, before writing, before law, before money, before machines, there was fire.

And with fire came the first great human heresy:

the idea that nature's conditions were negotiable.

PART I

The First Betrayals of Nature

The earliest revolutions did not begin by overthrowing governments or replacing institutions. They began by altering the conditions under which human beings experienced nature itself. In these first breaks, humanity learned that the world could be edited, symbolized, stabilized, and eventually organized.

These chapters move slowly on purpose. They begin before civilization in the familiar sense and before politics in its later forms, because the first thresholds were not primarily governmental. They were existential. What changed first was not law or empire, but the relationship between human beings and the conditions they inherited.

CHAPTER 1

Fire: When Nature Lost Its Monopoly

Before fire, night belonged to the world.

It belonged to cold, darkness, predators, uncertainty, and the long authority of things human beings could not negotiate with. When the sun disappeared, so did much of the human claim to agency. Vision narrowed. Movement slowed. Vulnerability thickened. The world beyond the body was no longer something to inhabit confidently, but something to endure until light returned. Night was not simply an absence of day. It was a condition of submission.

Imagine, for a moment, what darkness meant before controlled flame became ordinary. Not metaphorical darkness, but actual darkness: total, environmental, alive with sounds one could not identify and shapes one could not read. Beyond a few feet, the world dissolved. The body's own senses no longer guaranteed confidence. Heat dropped. Predators moved under cover that humans could not neutralize. The social group tightened physically because distance became dangerous. Rest was necessary, but it was not peaceful in the modern sense. It remained tied to exposure.

Then, at some point deep in prehistory, human beings began to alter that condition.

No single spark changed everything at once. The archaeological record is too fragmented, the timelines too debated, and the process too gradual for mythology of that kind. But whatever the exact sequence, one fact matters more than the dramatic version of it: human beings learned, eventually and decisively, not merely to witness fire, but to control it. And in doing so, they achieved something so foundational that it is difficult to see it clearly now. They did not just gain a tool. They altered the terms of reality.

Fire is often described as an early technology of survival, and it was. It provided warmth, protection, cooked food, hardened tools, and made certain landscapes more livable than they had been before. But survival language, while true, is too small for what happened. Fire was not merely useful. It was civilizationally philosophical. It was the first durable demonstration that the world's conditions were not fixed in the way they appeared to be fixed.

Darkness could be edited.

Cold could be held back.

Predators could be warned away.

Food could be transformed before it entered the body.

Night could be extended into something more social, more deliberate, more human.

A boundary had been breached.

This matters because human history is shaped not only by inventions, but by the expectations inventions create. Once human beings discover that one limit can be modified, they do not simply enjoy the modification. They begin to relate differently to limits themselves. Fire may have been the first great lesson in environmental dissatisfaction. It taught humans, however dimly at first, that reality was not only a thing to submit to. Under the right conditions, it was a thing to intervene in.

For most of the history of life, creatures adapted themselves to the environment. With fire, human beings began a long reversal: instead of only adapting themselves to the world, they began adapting the world to themselves. That reversal would become one of the master patterns of civilization. It would eventually reshape landscapes, climates, food systems, political networks, labor systems, media networks, and now perhaps cognition itself. But in its earliest form, it may have looked like nothing more than a group of humans sitting longer into the dark than any creatures had before.

The smallest revolutions are often the hardest to feel once they have won.

Around fire, the day changed shape. A human community with controlled flame was not simply warmer than one without it. It had more usable time. More visible time. More social time. The circle of firelight created a temporary territory against the surrounding dark. Within that territory, food could be prepared, stories could be told, tools could be repaired, warnings could be shared, and social bonds could thicken.

There is a difference between huddling in darkness under threat and sitting in a perimeter of light you have made yourself. In the first condition, the world dominates. In the second, human beings begin to experience a small but decisive form of local sovereignty.

That sovereignty likely changed cognition indirectly through routine. A firelit group is not simply safer. It can remain gathered. It can revisit events. It can coordinate. It can teach. It can ritualize. Fire did not create society, but it may have intensified one of its essential conditions: repeated gathering under managed circumstances rather than only under natural ones.

Here is one reason fire matters beyond utility. It may have changed not only what humans could physically do, but how they began to experience shared life. Sociality under controlled conditions is psychologically different from sociality under exposure. A boundary of light gives a group something more than heat. It gives a center.

It also creates one of civilization's earliest emotional templates: inside and outside. Inside the ring of fire, warmth, visibility, voice, and temporary order. Beyond it, danger, opacity, and the return of forces not yet governed. That distinction is not yet political in the later sense, but it is already architectural. Human beings are learning what it feels like to construct a world against the surrounding unknown.

And once a world can be constructed, it can also become expected.

Fire also reaches into the body. Cooked food is not the same as raw food. It asks less of the teeth, the jaw, the gut, and the hours otherwise required for chewing and digestion. The cooking hypothesis remains debated in its details, especially regarding chronology and the exact role of cooking in human evolutionary development, but the broader significance is hard to ignore: fire did not only defend the body from the outside. It changed the body's relationship to nourishment on the inside. It made matter more available to human purposes before that matter was even consumed.

This is a fundamental civilizational pattern in embryonic form: first transform the world, then transform the self through the transformed world.

Of course, the story is not archaeologically simple. Evidence for controlled fire becomes clearer in some places far later than popular imagination would prefer. There are disputes over when habitual use began, how widespread it was, and how directly it influenced later human developments. A serious history must keep those uncertainties visible. Yet uncertainty about timing does not diminish the deeper philosophical significance of the achievement. Whether gradual or punctuated, whether regionally uneven or widely shared, control of fire marks a threshold in the history of expectation.

There is one more dimension of fire that anticipates the emotional structure of every chapter to follow: the relationship between the keeper and the kept.

Fire must be tended. This is not a metaphor. It is a physical requirement. A fire that is not fed fuel dies. A fire that is not watched can escape and destroy. The keeper of fire enters into a relationship of continuous obligation, a relationship in which the tool does not merely serve the user but demands service in return.

This reciprocal obligation is the hidden cost of every powerful technology. The farmer who keeps livestock must feed, shelter, and protect the animals — daily, without fail, regardless of weather, mood, or competing demands. The city-dweller who depends on infrastructure must participate in the systems that maintain it, through taxation, labor, civic obligation. The smartphone user who depends on their device must charge it, update it, manage its storage, protect it from damage, and allocate attention to its demands. In each case, the tool that serves the user also binds the user to a regime of maintenance and attention that, over time, becomes invisible, experienced not as a cost but as the ordinary texture of life.

Fire established this template. The first fire-keepers were the first humans to discover that power comes packaged with obligation, and that obligation, once absorbed, stops feeling like obligation. It becomes routine. It becomes "just what you do." It becomes, in the language of this book, normal.

Every subsequent chapter describes a variation on this theme. Agriculture binds the farmer to the soil. The city binds the resident to the system. Writing binds the literate to the record. Money binds the participant to the abstraction. Industry binds the worker to the clock. Media binds the audience to the broadcast. The computer binds the user to the interface. The smartphone binds the person to the device. AI may bind the mind to the output.

The binding is never forced. It is always, at least initially, chosen. And the choice is always, at least initially, rational — the benefits of the technology genuinely exceed the costs of the obligation. But the obligation, once established, acquires its own inertia. The fire cannot easily be let go once the community depends on it. The field cannot easily be abandoned once the investment is made. The phone cannot easily be put down once the social expectations are set.

What emerges is the deepest pattern in the book: not that technology enslaves, but that technology entangles. And entanglement, unlike enslavement, does not feel like a loss of freedom. It feels like the way things work.

Nature had not been defeated.

But it had been challenged.

And with that challenge came one of civilization's oldest temptations: the temptation to treat every discomfort as a design flaw.

Fire solved real problems. It also created a new posture toward the world. Once warmth could be produced, the cold no longer appeared only as fate. Once darkness could be held back, night no longer appeared only as law. Once food could be cooked, rawness no longer appeared only as necessity. The very success of fire made natural conditions look less like permanent boundaries and more like circumstances awaiting revision.

In one sense, that is the beginning of human greatness.

In another, it is the beginning of human restlessness.

A solved friction does not remain merely solved. It changes the standard by which experience is judged. After fire, cold feels colder when it returns. Darkness feels less acceptable where light is possible. The removal of a burden becomes the creation of an expectation.

That sequence will repeat for the rest of this book.

A revolution removes a friction.

There is a scene worth constructing carefully, because it captures the moment when the pattern of this entire book first appears.

Imagine a band of *Homo erectus*, perhaps forty individuals, camped on a riverbank in East Africa, roughly five hundred thousand years ago. They have kept fire for generations. It is as natural to them as breathing. The children have never known a night without it. The elders cannot remember learning to tend it; the skill is as automatic as walking. The fire is simply there — part of the landscape, part of the rhythm of the day, part of what it means to be alive in this place.

Now imagine that one individual, sitting at the edge of the firelit circle, looks up at the stars and experiences something that no animal without fire could experience: leisure. Not the torpor of exhaustion. Not the stupor of satiation. But genuine cognitive leisure — a moment of consciousness free from immediate survival pressure, free from hunger, free from cold, free from the vigilance that darkness would otherwise demand. A moment in which the mind, released from the urgent, can attend to the interesting.

What does that mind think about? We cannot know. But we can observe that the conditions for abstract thought, for wondering, for imagining, for asking "what if?" — are precisely the conditions that fire creates: safety, warmth, social proximity, extended wakefulness, and freedom from the sensory tyranny of the dark. Fire did not merely extend the day. It created a new category of time — protected time, social time, reflective time — that would eventually become the medium in which language, myth, art, music, and every other form of symbolic culture would develop.

The anthropologist Polly Wiessner's research among the Ju/'hoansi of the Kalahari provides the closest modern analogy to what this firelit leisure might have looked like. Wiessner recorded conversations around daytime and nighttime fires over months of fieldwork. The daytime conversations were overwhelmingly practical: who would hunt where, how to resolve a simmering dispute, who owed what to whom. The nighttime conversations, extended by firelight into hours that would otherwise have been silent, were overwhelmingly narrative: stories about distant people, about journeys, about the spirit world, about events that had happened long ago or might happen in the future. The fire did not merely illuminate the night. It illuminated the imagination.

This is not a minor historical detail. It is the foundation of everything that follows. Without the extended social hours that fire created, the selective pressures that favored increasingly complex language may never have intensified. Without language, the symbolic culture that makes all subsequent revolutions possible could never have developed. Without symbolic culture, agriculture, cities, writing, money, and every other institution this book describes would be inconceivable. The chain of causation runs directly from the first campfire to the server farms that power modern artificial intelligence. It is a chain of unbroken dependency that spans the entire history of the species.

And at every link in that chain, the same pattern appears: a new capability is developed, it confers enormous advantages, it becomes indispensable, it reshapes the organism or the society that depends on it, and the reshape becomes invisible, absorbed into the baseline of ordinary life until it no longer registers as a change at all.

Fire is where that pattern begins. But it is not where the pattern ends. It never ends. It only accelerates.

The removal feels like liberation.

The liberation becomes expectation.

The expectation becomes baseline.

The baseline becomes invisible.

Then the next revolution begins.

Fire, then, was not only the first great tool. It was the first great normalization. It introduced humans to an edited condition of life and, in time, made that edited condition feel natural. No one born into a fire-using community would experience managed flame as an astonishing rebellion against the structure of the world. They would experience it as life. This is how revolutions disappear into culture. The most successful ones cease to feel revolutionary at all.

The effect is why fire deserves to be understood not simply as prehistory, but as prelude.

It is the first chapter in the long human practice of refusing to live entirely on nature's terms. It is the first successful argument against passive acceptance. It is the first durable proof that a discomfort can become a problem, a problem can become a technique, and a technique can become a way of life.

But fire did something else, something even more consequential.

It trained human beings to imagine that the world itself could be reorganized around their needs.

That imagination has a darker undertone than progress narratives usually admit. The creature who learns to soften winter with flame does not merely become safer. It becomes less patient with winter. The species that can delay darkness begins, slowly, to resent darkness itself. Invention does not only expand power. It alters emotional tolerance.

This matters because many later revolutions will not first transform capacity. They will first transform patience.

Once a people have lived inside an edited condition long enough, older conditions begin looking less like reality and more like failure. The inconvenience starts feeling unjust. The limit starts feeling outdated. The unmodified world starts feeling negligent. That emotional shift is one of civilization's oldest engines.

The modern mind did not invent dissatisfaction with friction. It inherited it from a very long lineage.

And fire is where that lineage becomes visible for the first time.

Once the imagination of revision exists, it does not remain confined to warmth and light. It reaches food, land, memory, time, value, labor, intimacy, and thought.

Civilization begins there.

Not with a flame alone, but with the expectation that the terms of existence are negotiable.

And once that expectation enters history, the next question is inevitable:

if the environment can be altered, can life itself be organized?

There is an experiment you can perform in your own mind to understand what fire meant. Go outside on a winter night with no phone, no flashlight. Stand in the dark. Feel the cold settle and the darkness press in. Feel how the body tenses, how the eyes strain against a blackness that gives nothing back. Now imagine a flame, just enough to push the dark back a few meters. Feel how the body relaxes, how the mind quiets. That shift, from siege to shelter, is the oldest human technology in action. It is so deeply encoded that most people feel it instantly.

Fire did not merely change the environment. It changed the emotional register of human existence. It introduced manufactured safety, the experience that security derived not from the absence of danger but from active intervention against it. The campfire does not eliminate the wolves. It holds them at bay. That holding, that active maintenance of a perimeter against the wild, is the template for every security system, every city wall, every firewall, every encryption protocol that would follow.

The philosopher Gaston Bachelard wrote that fire is the first object of reverie, the first occasion on which a conscious being sat before a phenomenon and simply watched, mesmerized, without practical purpose. The flickering of flame is hypnotic in a way few natural phenomena match: variable enough to hold attention, rhythmic enough to calm the mind. The campfire may be the first entertainment technology — the first object tended primarily for the experience it produced rather than any material output. Television, in many respects, is a domesticated campfire: a glowing screen in the living room that holds attention and creates a social center.

Fire also introduced catastrophic technological failure. A fire that escapes control can destroy everything it was meant to protect. The technology that provides safety also, under wrong conditions, provides total destruction. This duality, the same tool serving both creation and annihilation, would recur with every subsequent technology: agriculture can produce famine, cities can become death traps during plague, nuclear energy can power civilization or destroy it, and AI may enhance or displace human capability.

Fire created the first human experience of ambivalence toward a technology. The keepers of fire need it, depend on it for warmth, cooked food, and social life. Yet fire burns, escapes, consumes. A campfire not watched can destroy a camp. Fire disciplines its keepers even as it serves them. This ambivalence, needing what might hurt you, depending on what you cannot fully control, is the emotional signature of every revolution in this book.

Consider the archaeological specifics. At Gesher Benot Ya'aqov in Israel, excavated layers dating to approximately 790,000 years ago contain clusters of burned flint microartifacts concentrated in distinct spatial zones, evidence suggesting not random wildfire but repeated, deliberate burning in designated areas. The site's excavators argue that this pattern indicates spatial organization of fire use: certain areas of the living floor were used for burning while others were kept clear. If this interpretation holds, it suggests that the relationship between hominids and fire was already, three quarters of a million years ago, more than merely opportunistic. It was architectural.

The result is a word worth pausing on: architectural. Before any wall was built, before any structure was raised, fire created the first organized human space. The hearth, whether a ring of stones, a shallow pit, or simply a patch of ground repeatedly blackened by use — divided the world into zones. Here, warmth. There, cold. Here, safety. There, danger. Here, us. There, everything else. That binary — inside the circle and outside it — is the template for every built environment that would follow: the hut, the village, the city wall, the national border, the gated community, the password-protected network. All are descendants of the firelit circle.

The neuroscience of fire-watching adds another dimension. Studies of brain activity during campfire observation show decreased sympathetic nervous system activation — reduced heart rate, lowered blood pressure, diminished cortisol — consistent with a deep relaxation response. This is not merely anecdotal comfort. It appears to be a physiologically measurable reaction, possibly shaped by evolutionary exposure to controlled fire over hundreds of thousands of years. The human nervous system may be, in a literal sense, calibrated to respond to firelight as a signal of safety. If so, then fire did not merely change human behavior. It shaped the neurochemistry of relaxation itself.

This has implications beyond biology. If the human nervous system evolved partly in response to fire, then the species is not merely a fire-user but a fire-dependent organism at the physiological level. Remove fire from human evolutionary history, and you do not get the same animal. The brain is different. The gut is different. The social behavior is different. The circadian patterns are different. We are, as a species, an artifact of our oldest technology.

That thought is worth sitting with. Every subsequent chapter in this book describes a technology that changed human civilization. But fire may be the only technology that changed the human organism itself — not through genetic engineering or pharmaceutical intervention, but through the slow, patient sculpting of natural selection acting on populations that had incorporated an exosomatic energy source into their survival strategy. Fire did not merely give humanity an advantage. Over deep time, it became part of what humanity is.

And here is the final, most unsettling implication: if fire changed us, then every subsequent technology has the potential to do the same. Not necessarily through biological evolution — the timescales of modern technological change are far too short for that, but through the slower, subtler process of psychological and cultural adaptation that this book documents. Each revolution changes behavior first, then habits, then expectations, then values, then identity. The creature that emerges from the other side of a fully absorbed revolution is not the same creature that entered it. The change is not genetic. It is deeper than conscious. It is the rewriting of what feels normal.

Fire taught us that lesson first. We have been learning it ever since.

Fire was also, almost certainly, the first technology to create social inequality. The ability to start, maintain, and manage fire was a skill. Some individuals were better at it than others. And in a world where fire meant the difference between comfort and suffering, between safety and vulnerability, the skilled fire-keeper held a form of social power. This power was not formalized or institutionalized — that would come later, with agriculture and cities. But the seed was already present: the person who controlled access to a critical technology occupied a privileged position within the group.

This dynamic — technological skill translating into social power — would recur with every subsequent revolution. The scribe who controlled writing, the printer who controlled reproduction, the industrialist who controlled machinery, the programmer who controlled code, the platform that controls attention — all are descendants of the first fire-keeper, the first human who held a technology that others needed and could not easily replicate.

Fire gave humans the power to reshape their environment. But there was another revolution underway, slower, subtler, and in some ways more transformative. It did not change the physical world. It changed the world inside the mind.

It was language.

CHAPTER 2

Language: When Reality Became Shareable

Before human beings could build large worlds together, they first had to learn how to live inside the same invisible one.

That is what language made possible.

Long before cities, law, writing, money, bureaucracy, or artificial intelligence, there was already a quieter revolution underway: the transfer of reality from direct experience into shared symbol. A sound could now point to a thing. A thing could now carry an agreed meaning. A meaning could now survive the moment in which it was first needed. And once meanings could survive, they could be repeated. Once repeated, they could be taught. Once taught, they could govern.

Language is often described as a tool of communication, which is true in the same way that fire can be described as a source of warmth. The description is accurate, but too small. Language did not merely help humans exchange information. It allowed them to coordinate around things that did not physically exist in front of them. It allowed them to name absences, recall the dead, warn about unseen dangers, imagine future outcomes, assign roles, preserve taboos, elevate myths, and construct obligations that lived in the mind before they lived anywhere else.

With language, reality became portable.

That portability changed everything.

Once a word could reliably evoke not just an object but a rule, a memory, a promise, a role, or a threat, the human group no longer had to depend only on what was visible and present. It could begin to operate through what was collectively understood.

The moment a group can organize itself around shared meanings, the group gains an striking power. It can coordinate across time. It can preserve instruction. It can maintain invisible boundaries. It can punish not only harmful acts, but violations of categories. It can tell a child what kind of person they are before that child has done anything to become it. It can convert memory into identity and identity into obligation.

Language does not merely describe worlds. It installs them.

Here is one of the oldest and most decisive forms of human power. Physical force can move bodies. Language can shape the conditions under which bodies understand themselves. A spear may compel obedience in the moment. A story can compel it across generations. A command may govern behavior temporarily. A shared vocabulary can govern what a people experience as honorable, shameful, sacred, polluted, real, or impossible.

Before language reaches this level, humans confront danger primarily through perception. After language reaches this level, they also confront it through interpretation.

That distinction matters. A person who sees fire withdraws from heat. A person who is told that a certain place is cursed, a certain act is taboo, a certain ancestor watches, or a certain tribe is unclean is now living inside a reality structured not only by sensory evidence, but by transmitted meaning. Language expands the reach of intelligence, but it also expands the reach of fear, authority, memory, and imagination. It becomes the first technology through which one human mind can enter another without physical contact and leave behind architecture.

In that sense, language is the earliest infrastructure of civilization and the earliest infrastructure of control.

This does not make it sinister. It makes it foundational.

Without language, there is no durable kinship system, no multi-generational planning, no inherited law, no ritual order, no scalable trust, no mythic identity, no organized religion, no military coordination, no collective memory, no shared future. Language makes all of these possible because it gives human beings access to a reality beyond the immediate present. It turns thought outward and social life inward at the same time. It allows people to inhabit a world not only made of matter, but of meanings that exist because enough minds agree to carry them.

Here is one reason human beings are such strange creatures in the history of life. They do not live only in ecosystems. They live in symbolic environments.

A forest is real. A boundary is also real if enough people are willing to kill or die for it. A debt is real if a community enforces it. A marriage is real if a society recognizes it. A god is real in its consequences if enough minds organize around it. A nation is real in war even though no one has ever touched one with bare hands. The same species that eats, bleeds, shivers, and reproduces also lives inside constructs made of sound, repetition, memory, and belief.

Language is what makes that possible.

It is tempting to celebrate this as pure triumph, and in many ways it is. Language widened the human horizon beyond anything our ancestors could have achieved through gesture, instinct, or immediate perception alone. It enlarged cooperation, strengthened continuity, and allowed knowledge to accumulate. It likely helped early humans coordinate hunts, care for children, transmit practical skill, navigate danger, and preserve social order. The ability to tell another person not only what is here, but what was there, what might come, what must not happen, and what we all agree this means, is one of the most powerful evolutionary advantages ever achieved.

But every revolution that expands human power also expands human exposure.

There is another dimension of language that matters for this book and that is rarely discussed in popular accounts: language's relationship to time.

Before language, a creature lives in an eternal present. It can remember, memory predates language by hundreds of millions of years, but it cannot *discuss* its memories with others. It cannot say "last winter the river flooded and the animals moved to higher ground." It cannot say "if we collect extra berries now, we will not be hungry when the cold comes." It cannot say "my grandmother told me that her grandmother told her that the mountain once caught fire." These are all acts of temporal displacement, the ability to talk about the past, plan for the future, and transmit information across generations through purely symbolic means.

Language gave humanity access to time in a way no other species possesses. Through language, the past became a shared resource — a collective memory bank that any member of the group could access, contribute to, and draw upon. Through language, the future became a shared project, a domain of plans, predictions, promises, and contingencies that could be discussed, debated, and coordinated in advance of actual events.

What emerges is one of the foundations of cumulative culture — the ability to build knowledge across generations rather than starting fresh with each birth. A chimpanzee that discovers an efficient nut-cracking technique may teach it to its immediate companions through demonstration. But when that chimpanzee dies, the technique must be rediscovered or independently transmitted by observers. Language allows a different kind of transmission: "Hold the stone like this, not like that, because if you hold it wrong the nut shatters." The instruction can be verbal. It can be given in the absence of the actual nut and stone. It can include information about failures — "I tried the flat stone and it didn't work", that save the learner from repeating unsuccessful experiments. And it can be transmitted across generations: a grandmother teaching a grandchild a technique she learned from her own grandmother, with modifications accumulated over decades.

This is how human culture ratchets upward while other species' tool use remains relatively static. Language allows each generation to inherit not just the tools of the previous generation but the *knowledge* about those tools, the failures, the refinements, the contextual wisdom that makes the difference between competent and expert use. Without language, every generation reinvents the wheel. With language, every generation improves it.

The implications for this book's argument are direct. If language is what makes cumulative culture possible, then language is what makes every subsequent revolution possible. Agriculture depends on accumulated knowledge about seasons, soils, and plant behavior transmitted through speech. Cities depend on coordinated planning that requires linguistic communication. Writing is, literally, language made visual. Money is a symbolic system that derives its meaning from shared linguistic understanding. Every technology in this book rests, ultimately, on the foundation that language laid.

There is one final dimension of the language revolution that this chapter must address: the relationship between language and deception on a civilizational scale.

Language does not merely enable cooperation. It enables manipulation. The same capacity that allows one mind to share truth with another allows one mind to plant falsehood in another. And the falsehood, once planted, can be experienced by the recipient as truth, not because they are foolish, but because the mechanism of linguistic comprehension does not include a built-in fact-checker. The brain processes the content of a sentence before evaluating its truth. By the time skepticism kicks in, the meaning has already been absorbed.

What emerges is why propaganda works. This is why advertising works. This is why political rhetoric works. This is why misinformation spreads faster than correction. The brain is built to absorb language first and evaluate it second. And in many cases, when the message is emotionally compelling, when it confirms existing beliefs, when it arrives from a trusted source, the evaluation never happens. The absorption is the endpoint.

This vulnerability is not a design flaw. It is an inevitable consequence of a cognitive system optimized for speed over accuracy. In the evolutionary environment that produced language, small groups of closely related individuals with aligned interests — rapid linguistic absorption was adaptive. If a trusted band member said "danger," it was better to react first and evaluate later. The cost of delayed response to a genuine threat was far higher than the cost of occasional false alarm.

But in a modern information environment — where the "band member" has been replaced by a feed, a broadcast, an algorithm, or an AI, and where the interests of the message-sender may be radically misaligned with the interests of the receiver — this rapid-absorption mechanism becomes a vulnerability. The same cognitive architecture that allowed our ancestors to respond efficiently to warnings from trusted companions now allows advertisers, propagandists, political operatives, and recommendation systems to plant messages in minds that process them before skepticism engages.

Language gave humanity the power to share truth. It also gave humanity the vulnerability to absorb falsehood. The two capacities are inseparable — products of the same cognitive machinery, operating under different conditions. And the conditions of the modern world — saturated with messages from strangers, optimized for engagement rather than accuracy, delivered at speeds that prevent reflection, have tilted the balance decisively toward vulnerability.

Every subsequent media technology described in this book, writing, print, broadcast, digital networks, AI — has amplified this vulnerability by increasing the volume, speed, and persuasiveness of messages while reducing the time and cognitive resources available for evaluation. Language created the channel. Every subsequent technology widened it.

And if language made all of this possible, then the characteristics of language, its power to persuade, to deceive, to construct shared realities that are experienced as fact — are baked into the foundation of civilization itself. The AI that generates persuasive text is not introducing a new vulnerability into human life. It is exploiting a vulnerability that has existed since the first speaker discovered that arranged sounds could change what other people believed.

The same medium that allows care allows manipulation.

The same symbolic richness that allows culture allows propaganda.

The same shared story that binds a tribe can also blind it.

Language is the first infrastructure in which social reality can outgrow physical reality. And once that happens, human beings become vulnerable to worlds that are not false exactly, but constructed, exaggerated, sanctified, inherited, and defended beyond direct evidence. They become able to coordinate magnificently. They also become able to submit magnificently.

That is one of the book's central patterns appearing very early: every revolution creates a gain in capability and a gain in governability.

Language made human beings more cooperative, but also more shapeable.

A creature that can be frightened by a story is different from one that can only be frightened by a predator. A creature that can organize around an invisible hierarchy is different from one that only responds to brute dominance. A creature that can be told what is noble, impure, heroic, forbidden, chosen, or cursed can be moved by forces far subtler than hunger or immediate threat. Language made human beings less trapped by the present. It also made them more vulnerable to inherited scripts.

At what point does shared meaning become a shared enclosure?

There is a phenomenon in modern cognitive science that illuminates the power of language with startling directness: the experience of people who acquire language late.

Cases of language deprivation, individuals who, due to deafness, isolation, or extreme neglect, reach adolescence or adulthood without acquiring a first language, provide a natural experiment in what the mind looks like without the symbolic scaffolding that language provides. The findings are consistent and sobering. Without language, individuals struggle not only to communicate but to organize their own thoughts. They have difficulty with sequential reasoning, with planning beyond the immediate future, with understanding cause and effect across time, and with distinguishing between their own mental states and those of others. They can perceive the world, act in it, form attachments, and experience emotions. But they cannot, without language, construct the kind of structured, temporally extended, causally connected narrative of experience that most people take for granted as the basic texture of consciousness.

This suggests that language does not merely express thought. It partially constitutes it. The inner life of a language-using mind is qualitatively different from the inner life of a mind without language — not because the hardware is different, but because language provides the organizational architecture that makes certain kinds of thought possible. Abstract categories, temporal reasoning, hypothetical scenarios, moral principles, mathematical relationships, causal chains — all of these depend, to a significant degree, on the symbolic scaffolding that language provides.

If this is correct, and the converging evidence from neuroscience, psychology, and linguistics suggests that it is — then language is not merely a communication tool. It is a cognitive prosthetic. It is an externally transmitted technology that, once installed, fundamentally alters the capabilities of the mind that receives it. The child who acquires language is not merely learning to communicate. The child is being equipped with a cognitive architecture that will determine, for the rest of their life, what they can think, how they can reason, and what they can imagine.

This has a direct and unsettling implication for the later chapters of this book. If language, a purely symbolic apparatus transmitted through social interaction, can fundamentally reshape cognition, then other symbolic systems can do the same. The categories of a legal code reshape what counts as a crime. The metrics of an economic system reshape what counts as value. The algorithms of a digital platform reshape what counts as relevant. The outputs of an AI reshape what counts as competent thought. In each case, the symbolic system does not merely describe reality. It participates in constructing the cognitive and social reality its users inhabit.

Language was the first cognitive prosthetic. Writing was the second. Money was the third. Computers were the fourth. AI may be the fifth. Each one changes not merely what humans can do but how they experience the act of thinking itself. And each one, once absorbed, becomes so familiar that the change it introduced disappears from awareness.

There is one more dimension of language that merits attention before we move to agriculture: its role in the construction of the self.

Before language, there is no reason to believe that early hominids experienced anything like a continuous, narratively structured sense of personal identity. They experienced. They remembered. They felt. But the experience of being a self — of having a story that extends from a remembered past through a perceived present into an imagined future — requires the kind of temporal, narrative, and reflexive cognitive machinery that language provides. The self, in this view, is not a biological given. It is a linguistic construction, a story that the mind tells itself, using the tools that language provides, about who it is, where it came from, and where it is going.

This may sound abstract, but its practical implications are enormous. The experience of personal identity, of being someone with a name, a history, obligations, ambitions, and a projected future — is the psychological foundation on which all subsequent social institutions rest. Property requires a sense of self that persists over time ("this belongs to me, and it belonged to me yesterday, and it will belong to me tomorrow"). Law requires a sense of self that can be held accountable ("you committed this act, and you are the same person now as the person who committed it then"). Morality requires a sense of self that can evaluate its own behavior against standards ("I should not have done that, and I will try to do better"). All of these depend on the narratively structured, temporally extended sense of identity that language makes possible.

If AI eventually develops the ability to construct narratively structured accounts of its own operation — to say, in effect, "I did this because of that, and I plan to do this next" — it will be crossing a threshold that language first made possible for humans. Whether such accounts would constitute genuine selfhood or merely its simulation is one of the deepest philosophical questions the species may ever face. But the question itself is only possible because language first created the template of selfhood that AI might one day imitate.

The answer is not simple, because language always does both. It opens and confines. It allows expression and imposes categories. It creates memory and creates orthodoxy. It gives access to complexity while also freezing reality into names that can outlast their usefulness. To name something is often to understand it better. But it is also to place it inside a frame. Once enough frames accumulate, people cease to experience life only as raw encounter. They experience it through a preexisting architecture of terms.

That architecture can be beautiful. It can also be total.

A child is born into language long before they can judge it. The world is narrated to them before they can narrate it back. Certain things are normal, certain things shameful, certain things ours, certain things theirs. A mother tongue is not merely grammar and vocabulary. It is a map of emphasis. It teaches what is worth distinguishing, what is worth repeating, what deserves reverence, what deserves disgust, and how reality has already been sorted before the child arrives to meet it.

Here is not accidental background to civilization. It is civilization in its earliest intimate form.

Think about the difference between a child merely witnessing fear and a child being told what to fear. The first condition teaches reflex. The second teaches worldview. A people that can classify danger symbolically can also classify purity, rank, loyalty, destiny, and deviance. That is why language does not merely expand the communicative field. It expands moral reach. Whole populations can be coordinated around stories about origins, enemies, obligations, and the kind of world they believe themselves to inhabit.

The effect is also why ritual matters. Ritual is language made bodily. It turns repeated utterance into enacted reality. The chant, vow, oath, blessing, curse, taboo, mourning song, initiation speech, and battle cry all demonstrate that language does not only represent a world. It helps stage one. Repetition deepens legitimacy. What is spoken often enough begins feeling older than it is and more necessary than it may be.

Here is the beginning of durable culture, but also the beginning of durable enclosure.

Language gives a people continuity because it allows meanings to survive the death of individuals. That is its brilliance. But that same continuity means no generation begins entirely fresh. Human beings inherit not only land, tools, genes, and dangers. They inherit structures of interpretation. They inherit words that carry judgments, roles that carry expectations, and stories that carry obligations. Long before formal institutions appear, language already makes institutionality possible.

In that sense, writing will later scale what language first invents.

Before a law can be inscribed, it must be speakable.

Before a myth can be canonized, it must be narratable.

Before a hierarchy can be administered, it must be linguistically imaginable.

Before a civilization can be governed by symbols on tablets and screens, it must first become governable by sounds in the air.

Language is where that governability begins.

It is also where humanity first becomes able to live at a distance from immediate reality. This distance is not necessarily deception. In many ways it is the origin of human richness. It is what makes planning possible, mourning possible, complex morality possible, and history possible. But it also means that human beings no longer inhabit only what is. They inhabit what is said to be, what is remembered as having been, and what is expected to become. They become creatures suspended between reality and narrative.

That suspension is the birthplace of civilization.

And perhaps also the birthplace of dystopia.

Because once symbolic worlds become strong enough, people will sacrifice for them, kill for them, work for them, and surrender to them. A boundary can be invisible and still produce bloodshed. A title can be imaginary and still command obedience. A deity can be intangible and still rule a calendar, a legal code, or an empire. A market can be abstract and still reorder daily life. A machine can be statistical and still inherit trust. Human beings do not need physical chains to be organized powerfully. They need shared meanings with enough authority to feel real.

Language gave the species that possibility.

Which is why it deserves to be understood not as a mere communicative upgrade, but as the first major revolution in shareable reality.

Fire altered the external world around human beings. Language altered the internal world between them. Fire created managed conditions in the environment. Language created managed conditions in the mind. Around fire, people gained longer evenings and warmer bodies. Through language, they gained a common invisible terrain. A tribe could now be more than co-presence. It could be a story held together over time.

And once human beings can live inside a shared symbolic order, the next step becomes possible.

They can begin to organize life itself around those meanings.

Not just speak of land, but claim it.

Not just name seasons, but plan around them.

Not just remember abundance, but attempt to produce it.

Not just coordinate in motion, but settle in structure.

That next step is one of the most significant in history.

Agriculture.

Consider a thought experiment that reveals what language made possible. Imagine two bands of early humans facing the same threat: a predator near a shared water source. The languageless band can communicate danger to members who are present through alarm calls and body language. But the information cannot travel beyond physical boundaries. No plan can be articulated. No past experience can be verbally recalled.

The language-using band operates in a fundamentally different cognitive universe. A scout describes the predator's size, location, and direction. Someone recalls that a similar predator was driven off last season. A plan is formulated: at dawn, a group approaches from one direction while others wait with projectiles. The plan accounts for contingencies. It coordinates actions of individuals in different locations. It draws on memories of past events to predict future ones.

Multiply this advantage across thousands of generations and hundreds of thousands of encounters, with weather, food sources, migration routes, tool-making, social conflicts, and you begin to understand why language was not merely useful but civilizationally constitutive.

Language enabled myth, and myth is a structuring force of the first order. A creation story does not merely entertain. It answers existential questions every group must address: Where did we come from? Why are we here? What are the rules? A group sharing a creation myth shares a cognitive framework, common assumptions about reality that make coordination possible without continuous negotiation. The myth does social regulation invisibly, from the inside, by shaping what individuals experience as natural and morally required.

This is the mechanism by which language becomes an instrument of psychological architecture. Every language community lives inside a narrative framework it did not choose, received through acquisition, and experiences not as a framework but as reality itself. The child who grows up hearing that the ancestors are watching, that God sees everything, that the market is rational, is not merely learning information. The child is being installed with an operating system — default assumptions that will shape perception and behavior for life.

The philosopher Ludwig Wittgenstein wrote that the limits of one's language mean the limits of one's world. Research in linguistic relativity has produced evidence that language structure affects spatial reasoning, color perception, temporal cognition, and moral judgment. Language does not merely describe reality. It participates in constructing the reality its speakers experience.

This has implications reaching directly into the present. When a technology company describes its surveillance apparatus as a "personalization engine," it constructs a framework in which surveillance feels like service. When an AI company calls its product an "assistant," it frames a power relationship in the language of servitude. When a platform calls its users a "community," it borrows the warmth of a word implying mutual obligation and applies it to an extractive relationship. Language remains the oldest and most powerful technology of persuasion.

For hundreds of thousands of years, the combination of fire and language was sufficient. Groups remained mobile, following seasons and migrations. But then the climate shifted, certain plants concentrated in fertile valleys, and some groups began returning to the same places, season after season, tending what grew there.

Agriculture was coming. And with it, the first cage that felt like a home.

CHAPTER 3

Agriculture: When Survival Became Scheduling

If fire challenged the authority of nature and language made invisible worlds shareable, agriculture introduced a new possibility: that life itself could be organized in recurring patterns of yield, storage, and control.

Before agriculture, human existence was tied more tightly to movement, seasonality, and immediate ecological responsiveness. The foraging life was not idyllic, but it was differently exposed. It demanded attention to variation. It did not permit the same degree of accumulation, permanence, or concentrated administration that agriculture would eventually make possible. A mobile group could carry memory, tools, kinship, danger, and ritual across terrain, but it could not easily create the kinds of surplus that invite taxation, inheritance systems, and durable hierarchy.

To encounter a cultivated landscape as a habitual way of life rather than a temporary intervention would have meant entering a new moral geometry. The land was no longer simply there. It had been divided, worked, anticipated, and tied to future expectation. A field implied return. It implied waiting. It implied watchfulness. It implied that life would now unfold under a pattern that was not merely natural, but managed.

Agriculture changed that by making the future storable.

When human beings began to cultivate, they did not merely produce food. They began to discipline time. The calendar grew heavier. Seasons became obligations. Land became an asset. Surplus became strategy. Hunger became more manageable in some ways and more catastrophic in others. A failed hunt and a failed harvest are not psychologically identical. One belongs to uncertainty. The other belongs to a system.

The result is why agriculture deserves to be understood not only as the domestication of plants and animals, but as the domestication of expectations.

A field is a strange human invention. It is nature reorganized into rows. It is life made legible enough to plan against. It does not eliminate risk, but it creates a zone in which labor, patience, timing, and property begin to matter in new ways. The field teaches a moral lesson as much as a material one: stay, tend, wait, guard, repeat.

For the first time on a civilizational scale, survival became scheduling.

That scheduling is easy to underestimate because modern societies inherited it so completely. But it marks an enormous psychological shift. Foragers adapt to changing ecologies by movement, variation, and opportunistic knowledge. Agriculturalists begin adapting by repetition, settlement, prediction, and deferred reward. The future stops being only something one enters. It becomes something one prepares, stores, and protects.

The gains were enormous. Agriculture made larger populations possible. It enabled durable settlement, division of labor, surplus storage, inheritance, state formation, and eventually the long arch of what we call civilization. But this is precisely why it is such an important chapter in the normalization pattern. The gains were so transformative that the costs became easier to absorb.

Those costs were real. Agricultural life often narrowed diet diversity, increased labor intensity, deepened exposure to disease in settled populations, and linked people more heavily to territory, hierarchy, and defense. Surplus invited theft. Storage invited taxation. Land invited conflict. Permanence invited administration.

Agriculture solved some forms of precarity by installing other forms of dependence.

A wandering group may be vulnerable to uncertainty, but it is difficult to tax. A settled agricultural population can produce more, but it can also be counted, governed, conscripted, and enclosed. This is one reason agriculture matters so much to the argument of this book. It did not only increase human control over food. It increased the legibility of human beings to systems.

Once a crop can be counted, so can a household. Once a field can be mapped, so can an obligation. Once a village can be anchored, so can authority. Agriculture creates the material basis for larger-scale organization not only because it yields more food, but because it ties people to rhythms that institutions can eventually read and manage.

Grain is especially important here. Not all foods lend themselves equally to storage, transport, division, and taxation. Cereals can be counted, stockpiled, measured, and appropriated in ways many other food sources cannot. That makes them politically significant, not merely nutritionally consequential. The early agricultural world therefore matters not only because humans learned to grow food, but because some forms of food made humans more administratively visible than before.

The shift in human psychology is easy to miss because the modern world inherited it so thoroughly. Agricultural life teaches people to think in terms of delayed return, protected surplus, territorial continuity, and inherited responsibility. It extends human concern beyond the immediate horizon. It gives the future weight.

That may sound purely ennobling, but weight is not the same as freedom.

A future stored in grain is a future that must be guarded. A harvest planned months in advance is a life more vulnerable to interruption. A settled village may feel safer than movement, but it is also less mobile when danger comes. The field gives continuity and exposure in the same stroke.

The result is one of the recurring bargains of civilization: what grants stability also grants leverage to whatever organizes stability.

Agriculture also changes how people see land. Land is no longer merely habitat, route, terrain, or shared environment. It becomes productive territory. That shift is moral, political, and psychological all at once. To cultivate is to invest in location. To invest in location is to defend it. To defend it is to define insiders and outsiders more sharply. Property begins gaining emotional density.

The result is not just a new economy. It is a new kind of person.

The agricultural mind is patient, repetitive, anticipatory, anxious about conditions, protective of accumulation, and increasingly attached to continuity. It is not necessarily more wise than the foraging mind. It is differently arranged. It is organized around recurrence.

There is one more aspect of the agricultural revolution that connects directly to modern life: the invention of routine.

Foraging life was varied by nature. Each day brought different terrain, different weather, different food sources, different challenges. The forager's body and mind were constantly adapting to changing conditions — a mode of existence that neuroscientists would later recognize as cognitively stimulating, producing the kind of varied sensory and motor experience that maintains neural plasticity and cognitive flexibility.

Agricultural life was repetitive by necessity. The same field was tended day after day. The same crops were planted season after season. The same grinding, hoeing, irrigating, harvesting motions were performed until they became automatic. The farmer's day was organized around routine, a predictable sequence of tasks that could be performed efficiently precisely because they did not vary.

This invention of routine, of the repeated, predictable, structurally identical day, is one of agriculture's most consequential psychological innovations. It is also one of its most invisible, because routine, once established, ceases to feel like a choice. It feels like the structure of reality.

Modern life is built almost entirely on routine. The alarm clock. The commute. The work schedule. The meal times. The exercise routine. The evening ritual. The weekend pattern. The annual vacation cycle. Each of these is a descendant of agriculture's original invention: the substitution of the varied, responsive, adaptive day of the forager for the structured, predictable, repeatable day of the farmer.

And, as with every revolution in this book, the substitution carries both benefits and costs. The benefits are real: routine enables planning, coordination, reliability, and the kind of sustained effort that complex projects require. The costs are equally real but harder to see: routine narrows experience, reduces cognitive stimulation, promotes automaticity over awareness, and creates the conditions for the particular modern pathology that psychologists call "autopilot", the experience of moving through days without fully inhabiting them, performing sequences of actions so familiar that consciousness barely participates.

The smartphone has intensified this autopilot quality by filling the gaps between routines with a new routine: the reflexive check. The moment between tasks that once produced a brief experience of unstructured awareness — the walk between meetings, the pause before dinner, the minutes before sleep, is now filled with a scroll, a glance, a check. The routine has become seamless. The gaps have been closed. And the mind, which once had natural interruptions to its automaticity, now moves from one routine to another without ever encountering the uncomfortable, stimulating, potentially transformative experience of having nothing to do.

And once human beings become arranged around continuity, larger structures follow almost inevitably. Villages become more durable. Durable villages create pathways to towns. Towns create pathways to cities. Storage creates pathways to records. Surplus creates pathways to elites. Cultivation creates pathways to organized dependence.

And so agriculture is one of history's most consequential normalizations. It persuades human beings to accept discipline by making discipline materially rewarding. Repetition begins to feel responsible. Permanence begins to feel civilized. Order begins to feel morally elevated above uncertainty.

The rows in a field are not just rows of grain. They are rows of expectation. They teach repetition. They teach hierarchy between effort and yield. They teach attachment to place. They teach that the good life may require being more governable than before.

This is where one of the deepest ambiguities of progress becomes visible. Agriculture gave humanity more abundance. It also made humanity more schedulable. It expanded possibility while narrowing spontaneity. It increased control while enlarging the system within which control had to be maintained.

That is why agriculture remains one of the great hinge chapters of human history. It changes food, population, settlement, hierarchy, and time all at once. More importantly for this book, it shows how quickly discipline can become lovable when it arrives as security.

Once survival became scheduling, the next step was concentration.

Once people, surplus, and obligation begin to accumulate in one place, a new invention emerges almost by necessity.

There is another dimension of agriculture's psychological legacy that connects to the present: the invention of scarcity-anxiety as a permanent psychological condition.

Foragers, according to anthropological evidence, did not typically experience chronic anxiety about food supply. Marshall Sahlins famously described foraging peoples as "the original affluent society," arguing that their wants were few and their environment usually sufficient to meet them. Agriculture inverted this relationship. By tying survival to a single annual event, it created a chronic background anxiety with no equivalent in foraging life. The farmer worries about rain, about pests, about storage, about next year.

The modern worker who worries about job security, the investor who worries about volatility, the homeowner who worries about property values, the retiree who worries about savings are all experiencing versions of the same anxiety that agriculture first installed: the chronic awareness that current prosperity is conditional, that the future is uncertain, and that failure to prepare may prove catastrophic. The vocabulary has changed. The underlying emotional architecture has not.

The city.

There is a detail in the archaeological record that captures agriculture's hidden cost with startling clarity. At the site of Abu Hureyra in Syria, skeletal remains from the transition period show distinctive joint damage in women's bones. Toe joints, knees, and lower vertebrae show injuries consistent with hours of daily grinding, kneeling before stone querns, processing grain into flour with the same motion, day after day, year after year. These injuries are absent in pre-agricultural skeletons from the same site. The foragers show varied movement: walking, climbing, carrying, bending in multiple directions. The farmers' bodies tell a story of repetition.

This is not a footnote. It is a physical record of what the agricultural revolution did to the human body: it narrowed the range of movement, increased repetitive labor, and inscribed its demands directly into the skeleton. The forager's body was shaped by variety. The farmer's body was shaped by obligation. And the farmer did not experience this as oppression. She experienced it as life.

There is a dimension of the agricultural revolution that receives too little attention in popular history: its effect on the human experience of risk.

A forager's risk is immediate and diversified. If one food source fails, another can be sought. If the fishing is bad this week, the hunting or gathering may be better. The portfolio of survival strategies is broad, and the time horizon of risk is short, today, this week, this season. Catastrophic failure is possible (drought, epidemic, predator attack) but it tends to be acute rather than chronic.

A farmer's risk is deferred and concentrated. Everything depends on the harvest. Months of labor, entire seasons of planning and investment, converge on a single critical period when the crop either succeeds or fails. If it fails — because of drought, flood, locusts, blight, frost, or any of the hundred other disasters that can destroy a field, the consequence is not merely a bad day but a bad year. Starvation. Debt. Servitude. The loss of land that may have been in the family for generations.

This concentration of risk changed the human psychology of planning. The forager could afford to be relatively spontaneous, adapting to circumstances as they arose. The farmer had to think in seasons, sometimes in years. The farmer had to learn delayed gratification at a level foragers rarely required — planting now for a harvest months away, repairing irrigation channels in the dry season for the wet season to come, storing surplus against the possibility of future failure.

This temporal discipline, the capacity to subordinate present impulse to future reward, is often celebrated as one of agriculture's psychological gifts to civilization. And it is a genuine capacity, one that enables long-term planning, institutional building, and the kind of sustained effort that produces cities, libraries, and cathedrals. But it is also a psychological transformation that carries costs. The agricultural mind is a worried mind. It is preoccupied with weather, with storage, with the soundness of the granary, with the possibility that this year's harvest will not be enough. The anxiety of the farmer is not a personal neurosis. It is a structural consequence of living inside a system where everything depends on events that cannot be fully controlled and whose failure is catastrophic.

This anxiety would find institutional expression in the religious practices of agricultural civilizations, which are overwhelmingly concerned with fertility, rain, harvest, and the propitiation of natural forces. The gods of agricultural peoples are weather gods, grain gods, river gods, deities whose primary function is to ensure the success of the harvest and protect against the catastrophic concentration of agricultural risk. Prayer, sacrifice, and ritual are not mere superstitions in this context. They are technologies of psychological management, ways of addressing an anxiety that the agricultural architecture itself generates but cannot resolve.

The modern equivalents are insurance, futures markets, government subsidies, and weather forecasting. The anxiety is the same. Only the management strategies have changed.

Agriculture also reshaped the human relationship with death and inheritance. In a mobile foraging band, property is limited to what can be carried. When a member dies, their possessions can be redistributed or abandoned without significant economic consequence. In a farming community, death raises urgent questions that foragers never faced: Who inherits the land? Who inherits the stored surplus? Who inherits the debts? Who inherits the obligations to maintain the irrigation infrastructure, the granary, the boundary walls?

These questions required answers, and the answers — inheritance law, primogeniture, dowry systems, marriage contracts, became some of the most consequential institutional inventions in human history. They determined who held power, who accumulated wealth, who was bound by obligation, and who was free. They created the structural conditions for hereditary aristocracy, landed gentry, and the persistent concentration of wealth that characterized agricultural civilizations from Sumer to the European Middle Ages.

The family, as a legal and economic institution rather than merely a biological and emotional one, is in many ways a product of agriculture. Before farming, kinship was important but property was limited. After farming, kinship became the primary mechanism for transmitting the most important asset any family possessed: the land. Marriage became an economic contract as much as a social one — a means of consolidating holdings, securing alliances, and ensuring the continuity of productive assets across generations.

What emerges is one more dimension of the agricultural revolution's invisible legacy: the family as an economic unit, with all the consequences that follow, patriarchal authority justified by the need to control inheritance, restrictions on women's autonomy justified by the need to ensure legitimate succession, social hierarchies organized around land ownership rather than personal capability.

None of this was inevitable. It was structural, a consequence of the particular economic logic that agriculture introduced. But structural consequences, repeated across thousands of years and hundreds of civilizations, acquire the force of the inevitable. They become "the way things are." They become, in the language of this book, normalized.

Here lies the essential psychological mechanism this book traces. Not that people are forced into conditions they hate, but that conditions reshape what people learn to consider routine. The farmer grinding grain was not a slave. She was a participant in a framework that had become, within generations, the only way her community knew how to eat.

Consider what it meant, psychologically, for the first farming communities to experience crop failure.

A forager who finds no food in one direction walks in another. The setback is immediate and solvable. A farmer who watches drought destroy three months of labor has no equivalent recourse. The field cannot be moved. The invested time cannot be recovered. The community cannot simply resume foraging — the skills have atrophied, the landscape has been cleared, the population has grown beyond what gathering can support. The farmer's relationship with failure is fundamentally different from the forager's: it is catastrophic rather than incremental, total rather than partial, psychologically devastating rather than merely frustrating.

This experience of concentrated, irrecoverable failure is something new in human psychological history. Foragers experience setbacks. Farmers experience ruin. And ruin creates a psychological need that foragers did not possess: the need for insurance, for reserves, for protection against the catastrophic concentration of agricultural risk. This need drives the development of granary institutions, of social obligations to share surplus, of religious rituals designed to propitiate weather gods, and eventually of the formal insurance and financial instruments that characterize modern economies.

The psychological residue of agricultural risk, the ambient anxiety about the future, the compulsion to accumulate reserves, the fear that current prosperity is precarious, runs through modern economic behavior in ways that most people never consciously recognize. The person who saves compulsively, who hoards resources beyond any reasonable need, who experiences prosperity with an undercurrent of dread rather than simple enjoyment, is responding to a psychological template that agriculture installed ten thousand years ago. The granary is now a bank account. The weather god is now a financial advisor. The harvest anxiety is now portfolio anxiety. The underlying structure — the concentration of survival on a single vulnerable process, has not changed. Only the vocabulary has.

Agriculture also introduced the concept of waste, and with it, a new form of moral judgment. A forager who eats what the land provides and moves on leaves relatively little behind. A farmer who cultivates, harvests, processes, and stores grain generates a continuous stream of waste: chaff, spoiled stores, animal manure, human waste concentrated by sedentary living. The management of waste becomes, for the first time, a practical and moral concern. A clean granary is a sign of competence. A clean village is a sign of civic virtue. A failure to manage waste is not merely an inconvenience, it is a disease vector that can destroy the entire community.

The moral vocabulary of cleanliness, order, and discipline, concepts so deeply embedded in every subsequent civilization that they feel like universal human values — originates, in significant part, in the practical requirements of agricultural settlement. The farmer who does not maintain the irrigation channel endangers everyone. The household that does not manage its waste endangers the village. Discipline, orderliness, and the subordination of individual convenience to collective necessity become moral requirements rather than personal preferences.

The result is another dimension of agriculture's invisible legacy: the moralization of practical necessity. What begins as a survival requirement — keep the grain dry, maintain the channels, dispose of waste — becomes, over generations, a moral imperative that transcends its practical origins. Cleanliness becomes godliness. Order becomes virtue. Discipline becomes character. The practical logic of agricultural settlement is absorbed into the moral framework of the civilization that depends on it, until the moral framework feels independent of its practical origins, feels, in fact, like a timeless truth about the nature of human goodness.

Every subsequent civilization builds on this template. The industrial factory's demand for punctuality becomes the moral virtue of punctuality. The digital economy's demand for productivity becomes the moral virtue of productivity. The algorithmic platform's demand for engagement becomes, in subtle ways, the moral virtue of connectivity. In each case, the mechanism's practical requirements are absorbed into the culture's moral vocabulary until they feel not like requirements imposed by a system but like values inherent in human nature.

That absorption is the deepest form of the pattern this book traces. Not merely the normalization of behavior but the moralization of behavior, the point at which a system's demands stop feeling like demands and start feeling like duties.

Agriculture also transformed humanity's relationship with other species. Foraging societies lived in ecological partnership with the natural world, reading animal behavior, understanding plant cycles. Agriculture replaced this with something narrower and more instrumental. Animals became livestock — bred, confined, slaughtered according to human schedules. Plants became crops — selected for yield, planted in monocultures. Domestication was, from one angle, an impressive achievement. From another, it was the first large-scale application of a principle this book traces through every chapter: the conversion of autonomous living systems into components of a managed system.

The world we inhabit is not natural. It is farmed. And the farming happened so long ago, and became so total, that the farmed world feels like the natural one.

PART II

The Invention of Organized Human Life

The social consequences of agricultural settlement were even more far-reaching than the physical ones. Consider what happens when a mobile group becomes sedentary. In a foraging band, disputes can often be resolved through the simplest of mechanisms: separation. If two individuals or factions cannot coexist, one group can simply walk away, move to a different part of the territory, join another band, find new foraging grounds. The landscape is large, the groups are small, and the cost of fission is manageable.

Agriculture eliminates that option. A farming community is anchored by its investment, cleared land, planted crops, irrigation infrastructure, storage facilities, permanent dwellings. Walking away means abandoning not just a social situation but an economic one. The farmer who leaves loses not only their community but their livelihood. This creates an entirely new social dynamic: the necessity of conflict resolution within a fixed community. Disputes that could once have been resolved through departure must now be resolved through negotiation, adjudication, or suppression.

What emerges is where formal governance begins. Not as an abstract political philosophy, but as a practical response to the problem of people who cannot leave each other but cannot always agree. The village elder, the council, the chief, the priest — these are not inventions of political theory. They are solutions to a problem that agriculture created: how to maintain social order among people who are stuck together.

The emergence of hereditary leadership follows a similar practical logic. In a foraging band, leadership is often situational, the best hunter leads the hunt, the most experienced navigator leads the migration, the wisest elder mediates disputes. Authority is distributed and contextual. In a farming community, where surplus must be managed, labor must be coordinated, and disputes must be settled continuously, there is pressure toward centralized, permanent authority. Someone must decide when to plant, how to distribute water, who owes labor to whom, and what happens when obligations are not met. That someone gradually acquires institutional power — power that can be passed to heirs, reinforced by ritual, and defended by force.

Warfare, too, changes with agriculture. Foraging bands can fight, and do, but the incentives for sustained, organized violence are limited when there is little to steal and plenty of space to move. Agriculture changes the calculus fundamentally. A granary full of stored grain is worth fighting for. A cleared, irrigated field is worth defending. A population tied to its land cannot flee without losing everything. For the first time, warfare becomes economically rational in a sustained, systematic way. Armies, fortifications, tribute mechanisms, and conquest — all of these are consequences of a world in which wealth can be accumulated, stored, and stolen.

James Scott makes the provocative argument that early states were essentially protection rackets, organizations that extracted surplus from farming populations in exchange for security (often from threats that the state itself had created or exacerbated). Whether or not one accepts Scott's framing in its strongest form, the underlying dynamic is difficult to dispute: agriculture created the conditions for extraction by creating surplus, concentration, and immobility. A population that grows grain in fixed fields, stores it in centralized granaries, and cannot easily move is a population that can be taxed. And a population that can be taxed is a population that will be taxed.

The psychological dimension of this transformation is perhaps its most important aspect for this book's argument. The agricultural revolution did not merely change what people did. It changed what people expected. It introduced the concepts of property, inheritance, seasonal planning, deferred gratification, and institutional authority into the fabric of daily experience. A child born into a farming community learns, from infancy, that the world is organized around ownership, obligation, schedule, and hierarchy. These concepts feel natural, as natural as the seasons themselves — because they have been present for as long as anyone can remember. But they are not natural. They are agricultural. They are consequences of a revolution that happened ten thousand years ago and has been absorbing its own strangeness ever since.

The deepest irony of the agricultural revolution is that it created the conditions for civilization, for cities, writing, law, science, art, and every subsequent achievement of the human species, while simultaneously creating the conditions for servitude, war, famine, epidemic disease, ecological destruction, and institutionalized inequality. The same surplus that enabled the first libraries also enabled the first slave markets. The same settlement that produced the first temples also produced the first prisons. The same grain that fed the first cities also fattened the first tyrants.

That duality — creation and constraint arriving in the same package, is the signature of every revolution in this book. And it begins here, in the fields.

Once human beings could alter conditions, share symbolic realities, and stabilize food, a deeper transformation followed: organized life. This part traces the inventions that made large-scale coordination, memory, exchange, and administration increasingly durable.

The move from dispersed adaptation to organized continuity is one of the decisive turns in the book. Human beings do not merely solve immediate problems here. They begin building structures that can outlast individuals, store obligation, and widen the scale at which social order can be maintained.

CHAPTER 4

The City: When Strangers Became a Architecture

The city is one of humanity's most improbable achievements.

To wake in a city is to enter a functioning miracle one did not build and cannot personally maintain. Water arrives. Waste leaves. Food appears. Streets hold. Rules operate. Strangers move around one another under systems so normalized they are usually noticed only when they fail. It is a place where thousands or millions of people who do not know one another can eat, trade, move, sleep, work, and survive in close proximity without constant collapse into chaos.

That improbability is what makes the city so revealing.

A village can still run on memory. A clan can still run on kinship. A city cannot. A city requires a different logic. It requires infrastructures sturdy enough to coordinate strangers.

That is the real revolution of the city: not density alone, but organized interdependence among people who no longer rely on direct familiarity.

The earliest known urban settlements emerged independently in multiple regions. Mesopotamia, the Indus Valley, the Nile Delta, the Yellow River basin, Mesoamerica — suggesting that urbanization is not a cultural accident but a structural response to specific conditions of density, surplus, and social complexity.

Uruk, in southern Mesopotamia, may have held between forty thousand and eighty thousand inhabitants by roughly 3500 BCE. Archaeological excavations reveal a settlement organized around monumental temple complexes surrounded by dense residential neighborhoods, workshop areas, and centralized storage facilities. The temples functioned as economic hubs — managing the collection, storage, and redistribution of agricultural surplus. Temple administrators coordinated labor, maintained accounts through proto-writing systems, and allocated resources across a population far too large for any individual to oversee.

Walk through reconstructed Uruk in your mind. A potter carries finished vessels to a distribution point managed by temple staff. A weaver delivers textiles against a recorded labor obligation. A farmer from the surrounding fields brings grain to a central granary, where it is measured and stored. None of these individuals chose the system. They were born into it. The obligations they fulfill are conditions of urban citizenship, as invisible and binding as the air.

Lewis Mumford, in *The City in History*, describes the city's emergence as both a creative explosion and a new containment, concentrating skill, innovation, and ambition while simultaneously introducing new forms of administrative control and systemic fragility.

Consider Mohenjo-daro, in the Indus Valley. Archaeological evidence reveals standardized brick sizes in a consistent 1:2:4 ratio, sophisticated drainage with covered channels beneath the streets, public baths, and gridded layouts suggesting central coordination on an impressive scale, yet conspicuously little evidence of palatial or militaristic architecture. Whatever governed Mohenjo-daro operated through infrastructural discipline rather than visible autocracy. The city taught its residents how to live by shaping the spaces through which life moved. That insight is directly relevant to the later chapters of this book, where digital environments perform the same function through software architecture.

The city gathers bodies and disperses trust into systems.

To live in a city is to depend daily on people one will never meet and infrastructures one cannot individually maintain. Water arrives through hidden channels. Waste disappears through hidden channels. Food enters from distant fields. Rules are enforced by institutions rather than kin. Time becomes more public. Risk becomes more collective. Vulnerability becomes shared without becoming intimate.

There is one more aspect of urban life that deserves extended attention: the city's relationship to disease.

For most of human history before cities, infectious disease existed but could not sustain itself in small, mobile populations. A pathogen that infected a band of thirty foragers would either kill its hosts, be fought off by their immune systems, or run out of new victims, all relatively quickly. The band was too small and too mobile for epidemic disease to establish itself as a permanent feature of life.

Cities changed this calculus catastrophically. Dense populations living in close quarters, sharing water sources, generating concentrated waste, and interacting with domesticated animals created the conditions for sustained epidemic disease. Smallpox, measles, typhoid, cholera, plague, tuberculosis — the great killers of human history — are, in epidemiological terms, diseases of civilization. They require large, dense, sedentary populations to sustain transmission chains. They are, in a meaningful sense, the city's shadow.

The historian William McNeill argued in *Plagues and Peoples* that infectious disease has been one of the most consequential forces in human history — shaping the outcomes of wars, the fates of empires, and the trajectories of entire civilizations. The Black Death, which killed an estimated one-third to one-half of Europe's population in the fourteenth century, was not merely a medical catastrophe. It was a civilizational trauma that reshaped labor markets, religious attitudes, social hierarchies, and political institutions across an entire continent.

And yet cities survived. People continued to move to them. The advantages of urban life, economic opportunity, cultural richness, social mobility, access to services and institutions — were so compelling that people accepted the elevated disease risk as a cost of participation. This is, once again, the pattern of this book: the gains are vivid and immediate; the costs are diffuse and statistical. The merchant who moves to London for its markets does not think about cholera. The student who moves to Paris for its university does not calculate their probability of dying from plague. They experience the city's benefits directly and its risks abstractly — until the epidemic arrives, and then the abstraction becomes terrifyingly concrete.

The parallel to modern technology is direct. The user who adopts a new platform experiences its benefits immediately — connection, convenience, entertainment. The risks, data extraction, attention manipulation, psychological effects — are diffuse and statistical. They affect populations more than individuals, and they manifest gradually rather than dramatically. Like urban disease, they are costs that become visible mainly in retrospect, after the population has already committed itself to the apparatus that produces them.

Cities also introduced a new relationship between the individual and the collective that is relevant to the book's larger argument: the relationship of anonymous interdependence. In a village, interdependence is personal, you depend on specific people whose names and histories you know. In a city, interdependence is systemic, you depend on thousands of strangers whose labor sustains the infrastructure on which your life depends, but whose individual identities are irrelevant to the functioning of the system.

The result is one of the most psychologically significant transitions in human history, and it is one that modern people experience so constantly that they have lost the ability to find it remarkable. Every time you turn on a tap, you are relying on the competence of engineers, maintenance workers, water treatment specialists, and administrators whom you have never met and will never meet. Every time you eat a meal, you are at the end of a supply chain involving farmers, transporters, processors, distributors, and retailers spanning multiple states or countries. Every time you walk safely down a street, you are trusting that a complex system of laws, enforcement, urban design, and social convention will prevent the strangers around you from harming you.

The city is humanity's oldest and most successful experiment in organized trust among strangers. Every institution, every technology, every digital platform that mediates relationships between people who do not know each other personally is building on the template the city established thousands of years ago. The question this book asks — whether that template can sustain the scale and speed of digital and AI-mediated interaction — is, in a sense, the same question the first cities asked: can strangers cooperate at scale without destroying each other?

The city creates possibility by concentrating difference. Trade intensifies. Skill specialization deepens. Culture thickens. Innovation accelerates. The city can liberate individuals from the total moral visibility of smaller communities. It can make anonymity possible. It can widen ambition.

There is one more observation about cities that connects directly to the final chapters of this book: the city as a training system for system-trust.

A person who grows up in a city learns, from earliest childhood, to trust systems rather than persons for the provision of essential services. Water comes from the system. Food comes from the system. Safety comes from the machinery. Education comes from the infrastructure. Healthcare comes from the system. The specific humans who operate these systems are interchangeable, a different nurse, a different teacher, a different bus driver — but the systems themselves are presumed to be reliable.

This system-trust is one of the most important psychological prerequisites for the digital age. A person who has been trained by urban life to trust impersonal systems — to rely on infrastructure maintained by strangers, to accept services delivered through institutional channels, to believe that the apparatus will work even though they cannot personally verify its operation, is a person who is psychologically prepared to trust digital systems. The leap from trusting the water system to trusting the email system is not large. The leap from trusting institutional bureaucracy to trusting algorithmic decision-making is not large. The city prepared the mind for the computer, and the computer prepared the mind for the algorithm, and the algorithm is preparing the mind for AI.

This chain of prepared trust is one of the reasons the digital revolution was absorbed so quickly. People did not need to be convinced to trust digital organizations from scratch. They had been trained, by centuries of urban life, to trust impersonal, invisible, institutionally maintained structures as a basic condition of daily existence. The digital revolution did not create system-trust. It inherited it, from the city, from the bureaucracy, from the infrastructure that urban civilization had been building for five thousand years.

But anonymity is not freedom by itself. It is often exchanged for systemic dependence.

The city matters here not merely as a location, but as a psychological threshold. The city trains human beings to live among strangers through abstractions: law, money, role, address, office, market, schedule, record. It weakens older forms of social coherence while strengthening newer forms of institutional reliance.

To say that the city makes strangers into a system is not to condemn it. It is to name its underlying accomplishment. The city solves the problem of scale by making systems non-optional.

That has consequences. A village failure may be local. A city failure can become civilizational. A broken well in a village is serious. A broken water system in a city is existential. A city converts convenience into infrastructure and infrastructure into dependency.

This also alters belonging. In a smaller world, belonging is often inherited through direct relation. In a city, belonging must be mediated through category, district, class, trade, citizenship, or law. The city widens identity while thinning familiarity. It allows people to become socially legible through symbols rather than only through kinship.

The effect is one reason the city pairs so naturally with bureaucracy. Once a population becomes too large to be known personally, it must be known administratively.

Here again the pattern repeats. A friction is removed. One no longer needs to know everyone personally to survive. But another dependence enters. One must now trust arrangements to coordinate what memory no longer can.

The city transforms human life by concentrating possibility. It transforms human psychology by normalizing managed coexistence.

The city is where strangers stop being an exception and become the background condition of life.

This background condition has deep moral implications. In dense urban life, individuals are more likely to encounter one another as roles than as whole persons. The baker, the guard, the scribe, the merchant, the official, the laborer. Specialization increases competence and productivity, but it also encourages society to see people through function. Cities do not invent role-based identity, but they intensify it.

They also intensify vulnerability to invisible failure. In a city, daily life depends on systems whose functioning is mostly hidden when they work well. That hiddenness becomes one of urban civilization's defining traits. Success becomes ambient. Dependence becomes difficult to feel precisely because it is so well managed when intact.

This will be another recurring pattern in the book: architectures become strongest when they fade into the background of ordinary experience.

The city is an early masterpiece of that disappearing act. A resident may feel independent while depending on roads, walls, drains, storage, law, trade routes, labor chains, and institutions far beyond individual command. This is not hypocrisy. It is the ordinary psychology of apparatus life.

It is also a place where scale makes strangers morally necessary. The people around you are unknown, but their competence becomes part of your survival. The baker need not love you. The engineer need not know your name. The watchman need not belong to your lineage. Their role is enough. Cities therefore teach a difficult civilizational lesson: social order can persist without intimacy if systems become strong enough.

That lesson is productive, and dangerous.

There is one more lesson the city teaches that runs through every subsequent chapter: the lesson of invisible infrastructure.

A well-functioning city is, to its residents, invisible. The water arrives when you turn the tap. The lights come on when you flip the switch. The streets are passable. The food is available. The waste disappears. None of these services announce themselves. They operate in the background, silently and continuously, and they are noticed only when they fail, when the water stops, when the power goes out, when the garbage piles up, when the roads flood.

This invisibility is not an accident. It is the mark of success. Infrastructure that demands attention is infrastructure that is failing. The best plumbing is plumbing you never think about. The best electrical grid is one you never notice. The ideal state of infrastructure is transparent, present everywhere, felt nowhere.

But this transparency creates a psychological condition that is highly relevant to the later chapters of this book: the condition of invisible dependence. A city resident who turns on a tap and receives clean water is, in that moment, dependent on an engineering architecture of extraordinary complexity, treatment plants, pumping stations, pipe networks, quality testing, maintenance crews, regulatory agencies. They are dependent on all of this, and they are aware of none of it. Their dependence is absolute and their awareness of it is zero.

This is the template for every subsequent form of technological dependence this book describes. The internet user who searches for information is dependent on an infrastructure of servers, cables, protocols, and algorithms of extraordinary complexity. The smartphone user who opens an app is dependent on a supply chain spanning dozens of countries and hundreds of companies. The AI user who asks a question is dependent on a training process involving billions of data points, thousands of GPUs, and months of computational effort.

In each case, the dependency is absolute and the awareness is minimal. The user experiences convenience. The infrastructure that produces the convenience is invisible.

This invisible dependence has a specific psychological consequence: it produces the feeling of independence. The city resident who turns on the tap feels self-sufficient — they are, after all, providing for themselves. They got up, walked to the kitchen, turned the handle. The fact that their "self-sufficiency" depends on an invisible infrastructure maintained by thousands of strangers does not register. They feel autonomous because the system that constrains their autonomy is invisible.

Here is one of the deepest paradoxes of civilized life: the more dependent you become on systems you did not create and cannot control, the more independent you feel — because the systems are designed to be invisible, and invisible systems feel like nature. The tap feels like a natural source of water. The electricity feels like a natural property of the socket. The internet feels like a natural feature of the air. The AI feels like a natural extension of thought.

None of these things are natural. All of them are engineered. All of them could fail. And all of them create a form of dependence that is invisible precisely because it is so well-managed.

The city invented this condition. Every subsequent technology has deepened it. And the deepening continues.

There is one more aspect of urban civilization that deserves attention: the city's creation of the stranger as a permanent social category.

Before cities, the stranger was an anomaly, a rare encounter, potentially dangerous, certainly unfamiliar, requiring careful evaluation. The village's social world was defined by the known: known faces, known families, known histories, known obligations. The arrival of a stranger was an event, disruptive, interesting, potentially threatening, but always temporary. The stranger was someone who appeared, was dealt with, and either departed or was absorbed into the community of the known.

The city makes the stranger permanent. In a city, you are surrounded, at all times, by people you do not know and will never know. The stranger is not an anomaly. The stranger is the primary social reality. Your neighbors are strangers. Your fellow market-goers are strangers. The people who make your food, build your roads, enforce your laws, and dispose of your waste are strangers. The entire infrastructure of your daily life is maintained by strangers whose competence you must trust without any personal basis for that trust.

This represents a thorough rewiring of human social cognition. The human brain evolved for a social world in which every individual was known, in which reputation was based on direct observation, trust was built through repeated interaction, and social evaluation was grounded in personal experience. The city requires a fundamentally different mode of social cognition: evaluation based on category rather than acquaintance, trust based on role rather than relationship, judgment based on appearance, manner, and institutional affiliation rather than personal history.

Modern social psychology has demonstrated that this mode of cognition carries systematic biases. When people evaluate strangers, which is to say, when people operate in the default mode of urban social cognition, they rely heavily on categorical cues: appearance, accent, clothing, perceived social class, racial and ethnic markers, gender presentation. These cues trigger stereotypes, rapid, largely unconscious categorizations that serve as cognitive shortcuts for evaluating people about whom one has no personal information.

Stereotyping is not a moral failing in isolation. It is a cognitive adaptation to the problem of navigating a social world filled with strangers — the problem that the city created. The human brain was not designed for a world of anonymous masses. It was designed for a world of known individuals. Stereotyping is the brain's attempt to manage the information deficit that urban anonymity produces.

But the costs are enormous. Discrimination, prejudice, xenophobia, and the systematic dehumanization of out-groups are all, in part, consequences of a cognitive system designed for intimate communities being forced to operate in anonymous ones. The city's gift of anonymity and freedom comes packaged with the city's curse of categorical thinking and systematic bias.

This connection between urbanization and prejudice is rarely made in popular discussion, but it is fundamental to understanding the social pathologies of modern civilization. The racism, classism, and xenophobia that plague every urban society on Earth are not merely moral failings of individuals. They are structural consequences of a form of social organization, the city — that requires human beings to evaluate strangers at a rate and scale for which their cognitive architecture was never designed.

And the digital world has amplified this problem to a degree the physical city never could. Online, every interaction is with a stranger. Every comment, every post, every profile is from someone whose personal history you do not know and whose categorical membership is your primary basis for evaluation. The internet is a city without faces, a social environment that pushes categorical cognition to its extreme. The tribal hostility, the political polarization, the dehumanization of opponents that characterize online discourse are not aberrations. They are the predictable result of an environment that maximizes exposure to strangers while minimizing the cues — facial expression, tone of voice, physical presence — that moderate categorical judgment.

The city invented the problem of the stranger. The internet universalized it.

It makes large civilization possible. It also trains people to accept environments where legitimacy flows less from familiarity than from procedure. Once that becomes normal, a person's faith in daily life shifts from known people to durable arrangements. The system, not the elder, becomes the guarantor of continuity.

The hidden emotional revolution of urban life.

The city therefore marks a significant shift in how humans inhabit collective reality. They no longer survive mainly through familiar persons and remembered custom. They survive through organized environments that outscale personal knowledge.

That is the price and power of urban civilization.

And once social life depends on systems that must remember more than individuals can carry, memory itself must change.

Speech and custom are no longer enough. Obligation must survive distance, transaction, succession, and scale.

That requires a new invention.

Writing.

There is a phenomenon that the German sociologist Georg Simmel identified in 1903, writing about metropolitan life, that captures the city's psychological revolution with remarkable precision. The city-dweller, Simmel argued, develops a "blasé attitude", a protective emotional numbness in response to the constant stimulation of urban life. The city produces so many encounters, so many demands on attention, so many stimuli, that the individual learns to shield their inner life from the flood. They become reserved, calculating, intellectually rather than emotionally oriented.

This is not a moral failure. It is an adaptation, a way of surviving an environment that would be emotionally unbearable if one remained as open and responsive as a villager. Simmel's analysis describes with uncanny accuracy the psychological posture that modern people adopt toward digital life as well. The mechanism is identical: overwhelming input produces protective withdrawal.

Imagine also the matter of surveillance. A village is a surveillance society — everyone watches everyone, privacy is minimal. The city introduces anonymity, which has always attracted those seeking freedom from the village's moral gaze. But the city does not eliminate surveillance. It replaces personal surveillance with institutional surveillance. The village watches through gossip. The city watches through bureaucracy — records of residence, taxation, military service, property ownership, criminal history. The city creates the first administrative identity: the individual known not through relationship but through recorded information.

Every digital system that processes personal information today, every credit report, social media profile, algorithmic recommendation, government database, is a descendant of the first clay tablets on which city administrators recorded names, occupations, and tax obligations five thousand years ago. The technology has changed beyond recognition. The structure of the relationship has not.

Robin Dunbar's research suggests humans can maintain about 150 stable social relationships. The city blows past this threshold by orders of magnitude, not by expanding social cognition but by substituting institutional systems for personal knowledge. You do not need to know the baker. You need to know the bakery exists and the price system works. The entire relationship is mediated by role and convention. The person behind the counter could be replaced tomorrow and your breakfast would not change.

And once social life rests on arrangements that must remember more than any individual can carry, memory itself must change. Speech and custom are no longer sufficient. Obligation must survive distance, transaction, succession, and scale.

That requires a new invention.

Writing.

CHAPTER 5

Writing: When Memory Left the Body

There are few inventions more revered than writing.

Picture a dispute in a world before writing and the same dispute after inscription. In the first, memory remains contestable, social, embodied. In the second, the mark endures. The argument changes because permanence has entered the room. It seems like pure preservation, pure civilization, pure advancement. Writing lets words endure, laws travel, stories survive, instructions persist, and knowledge accumulate beyond the lifespan of individuals.

All of that is true.

But the earliest and most transformative uses of writing were not primarily literary. They were administrative.

That matters.

Writing did not first become world-changing because it made poetry possible. It became world-changing because it made memory scalable.

The work of Denise Schmandt-Besserat traces the origins of Mesopotamian writing to something remarkably mundane: clay tokens used for accounting. Beginning around 8000 BCE, small shaped objects, cones, spheres, disks, served as counters representing quantities of goods. Over millennia, these tokens were enclosed in clay envelopes, then impressed onto the surface so contents could be verified without breaking the seal. Eventually, the two-dimensional impressions replaced the tokens entirely. Writing, in this account, began as a receipt.

That origin matters enormously. Writing was designed to stabilize economic claims across time and distance — not to capture beauty or wisdom. In Sumer, the scribal schools, *edubba*, or "tablet houses", trained young men in cuneiform over years of rigorous study: forming wedge-shaped marks, copying standard lists, eventually composing administrative documents and literary texts. Literacy was not a birthright. It was an institutional resource.

Archaeologists have recovered tens of thousands of cuneiform tablets, administrative records, contracts, diplomatic correspondence, mathematical texts, literary works, astronomical observations, medical recipes, and school exercises. A city like Ur or Nippur could not have administered its trade, tax revenues, or legal disputes without the scribal infrastructure that processed, stored, and retrieved written records.

Consider what this meant for the debtor. In a pre-literate society, a debt exists in shared memory, contestable, softened by context, eventually forgotten. In a literate society, a debt exists on a tablet. It survives protest, outlives witnesses, and can be enforced by officials who never knew either party. Writing hardens obligation. It removes the social lubrication that oral negotiation once provided.

The Code of Hammurabi, inscribed on a basalt stele around 1754 BCE, did not merely document existing practice. It fixed it, establishing that law could be consulted rather than merely remembered, applied consistently rather than adapted case by case. This was beneficial for consistency. It was also a dramatic reduction in the flexibility of oral legal traditions.

Before writing, memory lived mainly in bodies, habits, rituals, and spoken repetition. Communities remembered through people. Once writing appears, communities gain a way to externalize memory into objects. The implications are enormous. A tax can now outlive the official who declared it. A debt can outlive the conversation in which it was incurred. A ruler can speak across absence. A law can remain identical in wording after the speaker is gone.

Writing does not merely preserve language. It stabilizes obligation.

What emerges is why it becomes such a powerful instrument of state formation. Grain records, ledgers, boundary claims, inventories, temple accounts, contracts, legal codes, decrees: the written world remembers what power needs remembered. The body forgets. The tablet does not.

Once memory leaves the body, institutions become less dependent on proximity. Administration becomes portable. Authority becomes durable.

This produces astonishing benefits. Writing enables philosophy, history, scripture, science, education, and literature. It widens access to ideas across time. It creates continuity where oral cultures must work harder to preserve exactness. But again, the revolution cannot be understood only through its liberating face.

Writing also changes the balance between the individual and the system.

Oral memory is powerful but local. Written memory can be centralized. Oral obligation often leaves room for social negotiation. Written obligation narrows ambiguity. The written world may remember more, but it often forgives less. Once a debt is inscribed, denial becomes harder. Once a rule is codified, flexibility can shrink.

Writing gives law permanence, but permanence is never neutral.

One of the most significant psychological changes writing introduces is this: what is written begins to feel more real than what is merely remembered. Documentation gains prestige. The record begins outranking the witness. A statement written down is no longer just speech. It becomes evidence.

There is a specific moment in the history of writing that captures its civilizational impact with particular clarity: the invention of the alphabet.

Earlier writing systems — cuneiform, hieroglyphics, Chinese characters, required the memorization of hundreds or thousands of distinct signs. This limited literacy to a professional class that could devote years to training. The alphabet — first developed by Semitic-speaking peoples in the second millennium BCE and later adapted by the Greeks, who added vowels to create the order that would become the foundation of Western literacy, reduced the number of signs to roughly two dozen.

This was a democratizing innovation of the first order. A writing system with twenty-six signs can be learned in months rather than years. Literacy becomes accessible not only to professional scribes but, in principle, to any member of the population with sufficient motivation and basic instruction. The alphabet did not immediately produce universal literacy — social, economic, and political barriers persisted for millennia. But it removed the technical barrier. For the first time, the cognitive prerequisites for literacy were within the reach of ordinary people.

The consequences were transformative. Alphabetic literacy enabled the development of Greek philosophy — a tradition of sustained, cumulative, written argument that would have been difficult to develop in a culture where literacy required years of specialized training. It enabled the spread of early Christianity, a religion whose sacred texts could be read, in principle, by any literate convert, without the mediation of a priestly class. It enabled the emergence of democratic governance in Athens — a system that required citizens to read laws, debate proposals, and participate in jury service.

The alphabet is, in this sense, one of history's great equalizers, a technology that lowered the barrier between the literate elite and the illiterate majority. But it also intensified the power of the written word by vastly expanding the population that could be reached by it. A proclamation written in cuneiform reached a tiny audience of trained scribes. A proclamation written in an alphabet could, in principle, reach anyone who had learned the twenty-six signs. The same technology that democratized literacy also amplified the reach of propaganda, demagoguery, and institutional control.

This dual character, simultaneously democratizing and amplifying, recurs in every media technology that follows. Print democratized access to books and amplified the reach of propaganda. The internet democratized publication and amplified the spread of misinformation. AI democratizes cognitive capability and amplifies the potential for manipulation. In each case, the technology lowers a barrier and raises a risk, and the barrier and the risk are inseparable consequences of the same innovation.

That shift reverberates across civilization. Religion canonizes. Law codifies. Trade formalizes. Inheritance stabilizes. Bureaucracy expands.

Writing is therefore not just the storage of language. It is the infrastructure of continuity.

There is a dimension of writing's revolution that connects it directly to the digital age and that deserves careful attention: the relationship between literacy and class.

In every early literate civilization, literacy was a privilege. The ability to read and write was not distributed democratically. It was concentrated in a professional class, scribes, priests, administrators, whose skills gave them asymmetric power over the illiterate majority. This was not merely a matter of convenience. It was a structural feature of how literate societies organized power.

Take the position of an illiterate farmer in ancient Mesopotamia. His debts are recorded on tablets he cannot read. His tax obligations are calculated by officials using records he cannot verify. His property boundaries are defined by documents he cannot interpret. His legal rights and obligations are inscribed in codes he cannot access without the mediation of literate professionals. He is, in a meaningful sense, governed by a system whose contents are invisible to him.

This is not merely historical. The same dynamic operates today in every society where essential systems, legal, financial, medical, technological — are mediated by specialized codes that most people cannot read. A modern citizen who signs a mortgage agreement is typically signing a document whose legal implications they do not fully understand. A patient who consents to a medical procedure is typically consenting on the basis of information filtered through the expertise of professionals whose knowledge they cannot verify. A user who accepts the terms of service of a digital platform is typically agreeing to conditions they have not read and could not fully interpret if they did.

In each case, the fundamental structure is the same as it was five thousand years ago in Sumer: an essential system is mediated by a code that most people cannot read, and the people who can read it hold asymmetric power over those who cannot. The code has changed — from cuneiform to legalese to programming languages to algorithmic logic, but the structural relationship between the literate and the illiterate has not.

The result is one of writing's most enduring legacies: the creation of information asymmetry as a permanent feature of civilized life. Every subsequent information technology, print, computing, networking, AI, has both reduced and recreated this asymmetry. Print made books available to millions but created a new gap between the literate and illiterate. Computing made data processing available to institutions but created a gap between those who understood the networks and those who merely used them. AI makes cognitive capabilities available to everyone but creates a gap between those who design the systems and those who depend on them.

The pattern is consistent: each new information technology democratizes access to the outputs of the previous one while creating a new tier of expertise that concentrates power in the hands of those who master the new technology. The scribes of Sumer have been succeeded by the jurists of Rome, the clerics of medieval Europe, the programmers of Silicon Valley, and the AI researchers of the present day. The technology changes. The structure of information power persists.

Writing also created something that would prove enormously far-reaching for the later chapters of this book: the concept of the permanent record. Before writing, human experience was ephemeral.

Events happened, were remembered for a time, and faded. Reputations rose and fell with the living memory of the community. Mistakes could be, if not forgotten, at least softened by the passage of time and the fallibility of human recollection.

Writing ended that ephemerality for certain kinds of information. A debt recorded on a tablet does not fade. A criminal sentence inscribed in a legal record does not soften. A boundary marker does not shift with the passage of years. Writing created permanence, and permanence created a new kind of power: the power of the archive.

The archive is a technology of memory that outlasts any individual memory. It allows institutions to accumulate information over decades and centuries. It allows patterns to be detected across time scales that no living memory could span. It allows the past to be consulted, verified, and — crucially, used as evidence in present disputes.

Every subsequent information technology has extended the reach and power of the archive. Paper archives gave way to printed archives. Printed archives gave way to digital databases. Digital databases gave way to the vast, interconnected, searchable, permanent data repositories that now contain more information about more people than any previous civilization could have imagined. The internet, in this sense, is the ultimate archive, a permanent, searchable, globally accessible record of everything that has been published, shared, uploaded, posted, or transmitted in digital form.

The consequences of this permanent record for individual privacy, for the ability to be forgotten, for the right to reinvent oneself, and for the relationship between past behavior and present identity are among the most urgent social questions of the digital age. But they are not new questions. They are questions that writing first posed five thousand years ago, when the first clay tablet made the first debt permanent.

And continuity is one of the deepest forms of power.

A spoken command dies if no one carries it. A written command can survive on its own terms. A rumor may mutate. A text can reproduce itself with much greater fidelity. A lineage can now claim continuity through records. An empire can now remember itself through archives.

There is one final observation about writing that connects it to the present moment: writing's relationship to identity and self-construction.

Before writing, identity was primarily performative, you were who you were in the presence of others. Your reputation existed in the memories and opinions of people who had directly observed your behavior. Identity was live, contextual, and continually renegotiated through face-to-face interaction.

Writing introduced the possibility of constructing identity at a distance. A letter presents a version of the self — edited, composed, deliberate. A legal document creates a formal identity, a name attached to obligations, properties, and social positions. A royal inscription constructs a political identity — the king as his subjects should see him, rather than as he necessarily is. Writing made it possible, for the first time, to separate the person from the persona — to create a version of the self that could circulate independently of the self's physical presence.

This capacity for textual self-construction has only intensified with each subsequent communication technology. The printing press allowed individuals to construct public identities through published works. The photograph allowed the construction of visual identity. Television allowed the construction of performed identity. The internet allowed the construction of digital identity, profiles, posts, curated timelines. Social media turned identity construction into a continuous, real-time, multi-platform performance. AI is now enabling the construction of identities that may not correspond to any living person at all.

At each step, the distance between the person and the persona has grown. At each step, the capacity for self-presentation has become more sophisticated, more deliberate, and more susceptible to manipulation. And at each step, the audience's ability to distinguish between authentic self-expression and constructed performance has become more difficult.

Writing began this process. Print accelerated it. Photography transformed it. Television amplified it. Social media universalized it. AI may complete it, by making the production of convincing personas so easy and so cheap that the concept of "authentic identity" loses its operational meaning.

The question of what identity means in a world where any identity can be fabricated — where text, image, voice, and video can all be generated by systems that have no identity of their own — is one of the deepest questions the digital age poses. And it is a question that writing first made possible, five thousand years ago, when the first scribe realized that the marks on clay could say whatever the maker wanted them to say, regardless of whether it was true.

The invisible gain is scale. The invisible cost is increasing subordination to what has been fixed.

In earlier settings, the body held knowledge and social life remained closer to direct relationship. Writing shifts authority outward. It creates a second memory, one that does not tire, age, or disappear as quickly as flesh. That second memory becomes a civilizational advantage. It also creates the possibility that people will be governed by records they did not author and may not fully understand.

This is a familiar pattern now. The system remembers more than the person.

Modern life takes this so much for granted that the original strangeness is hard to recover. But writing is an astonishing event in human history. It is the moment civilization first learns to put memory outside itself and then trust that external memory enough to organize life around it.

The practical uses mattered first because they solved persistent problems of scale. Counting grain, recording obligations, tracking exchange, preserving decrees, and maintaining continuity across death or distance were not glamorous functions. They were stabilizing ones. And stabilization is often the hidden engine of historical power. The society that can remember externally can coordinate beyond the ordinary lifespan and capacity of the individual body.

That means writing does more than preserve information. It redistributes authority.

Knowledge once held by elders, specialists, or ritual carriers begins competing with recorded form. The written text can travel farther than the memory-holder. It can also outlast them. This gives institutions a new kind of endurance and gives future generations access to past commands in a way that oral societies mediate differently.

There is beauty in that. There is danger in it too. Once authority can survive in fixed form, frameworks can become more coherent and more rigid at the same time.

Writing also alters shame. A spoken accusation can fade or be renegotiated. A written accusation can remain, circulate, and reappear. A spoken contract depends heavily on community memory. A written contract can be consulted without community warmth. In this sense, writing not only strengthens law and administration. It cools social life. It makes agreement more durable, and often more impersonal.

That impersonality is not always a loss. It can protect fairness and continuity. But it also widens the distance between human intention and institutional consequence. Once something is recorded, structures can act on the record even after the living context has changed.

That is why writing belongs in the long history of normalization. At first, externalized memory would have been an astonishing intervention in human affairs. In time, it became so natural that modern people struggle to imagine social life without it. That is precisely what successful revolutions do. They become conditions of intelligibility.

The scale of the scribal enterprise reveals how quickly civilization became dependent on externalized memory. Archaeologists have recovered tens of thousands of cuneiform tablets from Mesopotamian sites — administrative records, legal contracts, diplomatic correspondence, mathematical texts, literary works, astronomical observations, medical recipes, school exercises. The sheer volume testifies to a civilization that had become, within centuries of writing's invention, structurally dependent on external memory for its most critical functions.

Consider what a city like Ur required in terms of written documentation. Trade agreements with distant cities. Records of grain stored in temple granaries. Lists of workers assigned to canal maintenance. Inventories of livestock. Legal rulings in property disputes. Marriage contracts. Loan agreements with specified interest rates. Royal decrees. Religious rituals. Mathematical tables for land surveying. Astronomical observations for calendar-keeping. Each of these functions depended on writing, and on the scribal class that maintained it.

The loss of written records could be catastrophic. When a city was sacked and its archive destroyed, the institutional memory of the community was destroyed with it. Debts could not be verified. Boundaries could not be confirmed. Legal precedents vanished. The continuity of administration, the thread that held complex urban life together, snapped. This vulnerability reveals something important: writing did not merely supplement human memory. It replaced it for certain critical functions. And once the replacement was complete, the institution could no longer function without the external system.

This pattern, supplementation becoming replacement becoming dependency, recurs throughout the book. It is the same trajectory followed by agriculture (supplementing foraging, then replacing it), by money (supplementing direct exchange, then replacing it), by computers (supplementing human calculation, then replacing it), and by AI (supplementing human cognitive labor, and beginning to replace it). In each case, the tool that was adopted for its convenience gradually becomes indispensable, and the capacity it was meant to assist gradually atrophies.

Writing also introduced a new relationship between the living and the dead. In an oral culture, the dead speak only through the memories of the living. When the last person who remembers a deceased elder dies, the elder's words are gone. In a literate culture, the dead can speak indefinitely. A law written by a king three centuries ago can still be enforced. A contract signed by a merchant who died decades ago can still be binding on his descendants. A religious text composed by prophets long dead can still command the behavior of millions of living people.

This is an unusual power, and it cuts in multiple directions. On one hand, it enables continuity, the preservation of wisdom, law, and institutional knowledge across generations. On the other hand, it enables a kind of temporal tyranny: the present can be governed by the decisions of the dead. A legal code written in a different world, under different conditions, for different purposes, can persist long after the circumstances that justified it have vanished. A religious text composed in one cultural context can be enforced in another. The dead hand of the past, preserved in writing, can reach across centuries to constrain the living.

This tension between continuity and constraint is built into the structure of literate civilization. Every written constitution, every legal code, every religious scripture, every institutional charter represents a decision that someone, at some point, inscribed in permanent form — and that subsequent generations must either obey, reinterpret, or formally overturn. The decision to write something down is, in a sense, a decision to give it power over the future. And the more thoroughly a civilization depends on written records, the more thoroughly it is governed by the decisions of people who are no longer alive to explain, modify, or retract what they wrote.

The result is one reason that revolutions, political, intellectual, religious — so often begin with the reinterpretation or rejection of authoritative texts. The Reformation was, at its core, a dispute about how to read the Bible. The American Revolution was, in part, a dispute about the meaning of English constitutional traditions. The scientific revolution was a systematic challenge to the authority of inherited texts by empirical observation. In each case, the revolutionary force was directed not against the practice of writing itself but against the accumulated authority of what had been written. The revolution sought not to destroy literacy but to liberate the living from the dead.

Writing gives memory immortality. But immortal memory can become immortal authority. And immortal authority can become invisible — so deeply embedded in the structure of daily life that it no longer appears as authority at all. It appears as how things are.

Once memory can be stabilized and exchanged, then value can be as well.

Money follows naturally.

CHAPTER 6

Money: When Value Stopped Looking Human

Money is so ordinary now that it seems almost natural.

Imagine two strangers meeting with no kinship, no shared ritual, no deep reciprocal history, and still being able to transact smoothly because both trust an abstraction. That is not a minor social convenience. It is one of civilization's great cognitive triumphs and one of its great moral simplifications.

Money is one of the most radical abstractions human beings ever normalized.

David Graeber, in *Debt: The First 5,000 Years*, dismantled the textbook narrative in which barter preceded money. Early economies, he argued, typically functioned through complex systems of credit, gift exchange, and social obligation. Money emerged not from barter but from the formalization of debts that had existed in more personal forms for millennia.

If money's origins lie in debt rather than exchange, then its deepest function was never simply to make trade easier. It was to make obligation portable and enforceable beyond the boundaries of personal relationship.

Look at the cowrie shell. In parts of Africa, South Asia, and Oceania, cowries served as currency for centuries, portable, durable, countable, difficult to counterfeit. They also participated, with terrible efficiency, in the slave trade. Human beings were bought and sold for quantities of shells, their lives reduced to numerical transactions. The shells did not create slavery. But they made the commerce of human lives calculable — and calculability embeds atrocity more deeply into economic logic.

The invention of coinage in Lydia around 600 BCE added standardization: guaranteed weight and purity, stamped with state authority, circulating among strangers without requiring personal trust.

The coin declared: you do not need to know me. The state guarantees this metal. That transfer of trust from person to institution is one of the recurring moves in administered life — appearing again with paper currency, banking, electronic payments, and cryptocurrency.

Georg Simmel, in *The Philosophy of Money*, argued that monetization transforms not only exchange but perception itself. Once enough of life can be translated into units, the imagination starts speaking in units. Price can feel firmer than context. Quantification can feel more trustworthy than memory. Abstraction begins to masquerade as neutrality.

At its core, money does something philosophically astonishing: it allows unlike things to be compared on a shared scale. Grain, labor, land, debt, metal, time, risk, and eventually nearly every object or service can be drawn into common equivalence. Money does not erase difference. It renders difference transactable.

That transformation changes more than markets. It changes perception.

In a world of direct exchange, value remains thick with context. The thing exchanged is tied to relationship, scarcity, status, trust, and circumstance. In a monetized world, value becomes portable. It can move farther from the social setting in which it originated. That portability is immensely liberating. It allows strangers to trade. It allows labor to be priced. It allows wealth to be stored, transferred, calculated, and accumulated at scale.

But every abstraction that enlarges freedom also risks thinning moral texture.

Money teaches societies to compare unlike things as though they belong on one scale. Once that habit becomes normal, the imagination of exchange expands dramatically. The question shifts from whether something is comparable to how much it is worth.

Here is one of the deepest rewrites in human cognition. It does not merely create commerce. It teaches equivalence.

When value stops looking human, flexibility rises. So does impersonality.

Money frees people from certain local dependencies. One does not need deep reciprocity with every person one trades with. That matters. Yet the same liberation means value can increasingly detach from relationship, memory, and local obligation. Money scales trust by replacing some of its old conditions.

The history of money is also a history of crises, and those crises reveal something important about the relationship between abstraction and trust.

Observe the phenomenon of the bank run. A bank holds deposits from thousands of customers but keeps only a fraction of those deposits in liquid form — the rest is lent out or invested. As long as customers trust that their money is available when they need it, the apparatus works. But if trust falters, if enough customers begin to doubt the bank's solvency, a feedback loop develops. Customers rush to withdraw their deposits. The bank, unable to convert illiquid assets to cash fast enough, fails. The failure confirms the fear that caused the withdrawal. Trust collapses, and the abstraction on which the entire system depended, the shared belief that the money in the bank is "real" — shatters.

The result is not merely a financial phenomenon. It is an epistemological one. A bank run reveals that money, in its modern form, is not a thing but a consensus. The numbers in your bank account are not backed by gold bars in a vault (they haven't been for decades). They are backed by the collective belief that the banking system is sound — a belief that is usually well-founded but is, at bottom, a matter of trust rather than physical reality.

The same structure of consensual reality applies to every form of money, from the earliest cowrie shells to contemporary cryptocurrency. The value of a dollar bill derives not from the paper it is printed on (which is worth fractions of a cent) but from the collective agreement to treat it as valuable. The value of a Bitcoin derives not from any physical asset but from the collective agreement of its users to accept it as a store of value and medium of exchange. In both cases, the "reality" of the money is entirely social — a shared fiction maintained by coordinated belief.

This is relevant to the book's larger argument because it illustrates a general principle: once a civilization builds its essential infrastructure on shared abstractions, the maintenance of those abstractions becomes a matter of civilizational survival. A society that loses faith in its money does not merely experience economic inconvenience. It experiences civilizational crisis, as happened in Weimar Germany, in Zimbabwe, in Venezuela, and in every other case of hyperinflationary collapse. The abstraction that made complex economic coordination possible becomes, when it fails, the source of catastrophic social disintegration.

The same principle applies, with increasing urgency, to every subsequent abstraction this book traces. Writing depends on the shared belief that recorded information is accurate and authoritative. Law depends on the shared belief that rules will be enforced consistently. Science depends on the shared belief that method produces reliable knowledge. Digital infrastructures depend on the shared belief that data is trustworthy and algorithms are fair. Each of these abstractions is immensely powerful as long as the belief that sustains it remains intact. Each becomes dangerous when that belief erodes.

Money teaches this lesson first: that the most powerful mechanisms in human civilization are also the most fragile, because they are built not on physical foundations but on psychological ones. A bridge fails when the steel gives way. A monetary system fails when the belief gives way. And belief, unlike steel, can collapse in an instant.

And so financial crises recur with such stubborn regularity despite centuries of institutional learning. The problem is not technical. It is structural: any machinery built on collective belief is vulnerable to collective doubt. And collective doubt, once it begins, is almost impossible to stop, because each act of doubt (each withdrawal, each sell-off, each refusal to lend) provides evidence that further justifies doubt.

The pattern is identical to the erosion of trust in any shared apparatus, including the digital systems that increasingly mediate contemporary life. When trust in a platform erodes (as it did with Facebook after the Cambridge Analytica scandal, or with Twitter after its acquisition and transformation), the effects cascade in the same way: users leave, content quality declines, remaining users leave faster, and the platform's value — which was always based on collective participation rather than intrinsic worth — evaporates.

There is one more scene worth constructing: the moment when money first creates the experience of abstract wealth.

Imagine a merchant in ancient Lydia, around 600 BCE, holding in his hand a small electrum coin, a standardized disk of gold-silver alloy, stamped with the seal of the king. The coin weighs a specific amount. Its composition is guaranteed by the state. It can be exchanged for goods anywhere in the kingdom, and increasingly beyond it, without the need to weigh, test, or negotiate the value of the metal.

The merchant holds in his hand something that no previous generation has possessed: portable, standardized, state-guaranteed abstract value. He can carry his wealth, not as grain, which rots; not as livestock, which must be fed and guarded; not as land, which cannot be moved, but as a small, durable, universally accepted token that represents value without embodying it.

This is an experience of abstraction so serious that it is difficult, from inside a fully monetized civilization, to appreciate its novelty. The coin is not valuable because of what it is. It is valuable because of what it represents. It is a symbol, a physical object whose significance lies entirely in the shared agreement of the humans who use it. Remove the agreement, and the coin is just a piece of metal. Maintain the agreement, and the coin is power, freedom, security, possibility.

The merchant who holds this coin is holding something new in the world: a piece of reality that is simultaneously physical and fictional. The metal is real. The value is agreed-upon. The coin exists in both worlds at once, the world of objects and the world of meaning. It is, in this sense, a descendant of language (which also creates shared fictions that are experienced as real) and an ancestor of every subsequent financial instrument (paper money, bonds, stocks, derivatives, cryptocurrency) that pushes the abstraction further from the physical.

The trajectory is clear: from grain (fully physical) to metal (partially physical) to coin (standardized physical) to paper (symbolic physical) to ledger entry (institutional abstraction) to digital record (purely informational) to cryptocurrency (mathematically defined). At each step, the physical component diminishes and the fictional component grows. At each step, the abstraction becomes more powerful and more fragile — more capable of coordinating activity at scale, and more dependent on the collective belief that sustains it.

Money's history is, in this sense, a rehearsal for the digital age, a five-thousand-year experiment in the construction, maintenance, and occasional collapse of shared fictions that are experienced as reality. The person who trusts a digital payment is the distant descendant of the Lydian merchant who trusted a coin. The mechanism of trust has changed. The structure of the fiction has not.

Money was the first great lesson in the power and fragility of shared abstraction. It will not be the last.

This is why money pairs so well with large states, urban life, taxation, recordkeeping, and empire. It is one of the great accelerants of scale because it compresses complexity into units.

And once value is unitized, life begins to be seen through measures that feel objective even when they conceal moral choices.

Wage labor grows easier to imagine in a monetary world. Debt becomes more legible. Trade expands beyond intimacy. Wealth becomes more comparable than ever before.

What is gained is flexibility, portability, and power.

There is a thought experiment that reveals the strangeness of money more effectively than any economic theory.

Imagine explaining money to an intelligent alien — a being that understands physics, biology, and social behavior but has never encountered a monetary mechanism. You would have to explain that certain objects, pieces of metal, slips of paper, entries in electronic databases, have no intrinsic value whatsoever but are treated by human beings as though they were the most valuable things in existence. You would have to explain that people will work for years to accumulate these objects, that societies will go to war over them, that individuals will betray friends and family to acquire them, and that the entire global economy — the production, distribution, and consumption of every good and service on the planet — is organized around their movement.

The alien would, reasonably, ask: why? What makes these objects valuable?

And you would have to give an answer that, stated plainly, sounds like a description of a collective hallucination: they are valuable because everyone agrees they are valuable. The agreement is self-reinforcing — the more people accept the fiction, the more useful it becomes, and the more useful it becomes, the more people accept it. But the foundation is not physical reality. It is shared belief.

Here is not a criticism of money. It is a description of its nature. And the description is relevant to this book because it illustrates a general principle: some of civilization's most powerful institutions are built not on material foundations but on psychological ones. Money works because people believe in it. Law works because people accept its authority. Science works because people trust its methods. Democracy works because people agree to abide by its outcomes. Remove the belief, and the institution collapses, not because the physical world has changed, but because the psychological world has.

This vulnerability — the dependence of civilizational infrastructure on sustained collective belief, is one of the recurring themes of the later chapters. The internet works because people trust that the information it provides is at least approximately reliable. Social media works because people believe that the connections it facilitates are at least approximately genuine. AI works because people trust that the outputs it generates are at least approximately competent. In each case, the system's power depends on a psychological foundation that is robust but not indestructible.

Money provides the template for understanding this dynamic because it has been around longest. The history of money is, in part, a history of the periodic collapse and reconstruction of shared financial belief, from the debasement of Roman coinage to the hyperinflationary crises of the twentieth century to the cryptocurrency volatility of the twenty-first. Each collapse reveals, with painful clarity, that the value people attributed to their money was always a shared agreement rather than a physical fact. And each reconstruction demonstrates that the agreement can be rebuilt — but only if enough people are willing to believe again.

The deepest lesson of money, for this book's purposes, is not economic. It is psychological. Money teaches that human beings are capable of constructing shared fictions of unusual power, fictions that can coordinate the behavior of billions of people across the entire planet. It also teaches that those fictions, however powerful, rest on a foundation of trust that can never be taken for granted.

Every subsequent chapter in this book describes a system that depends on a similar foundation. And every such mechanism faces the same fundamental vulnerability: the gap between the system's power and the fragility of the belief that sustains it.

What is lost, at least in part, is the resistance created by context.

Money does not create greed, but it gives greed remarkable efficiency. It does not create hierarchy, but it gives hierarchy mobility. It does not invent markets alone, but it turns markets into a language civilization can increasingly speak fluently.

What matters is not an anti-money argument. It is a reality argument. Money works because abstraction works. But abstraction always alters what feels visible.

The person who owes and the number that represents the debt are not emotionally identical. The worker and the wage are not morally identical. The land and the price of the land are not existentially identical. Yet monetized civilization encourages us to move quickly between those levels until the translation feels seamless.

There is one more dimension of money's revolution that connects directly to the AI era: the concept of quantification as a mode of understanding.

Money did not merely create a medium of exchange. It created a cognitive habit — the habit of translating qualitative experience into quantitative terms. This habit, once established, proved to be extraordinarily contagious. It spread from commerce to governance (taxation, budgeting, cost-benefit analysis), from governance to science (measurement, statistics, mathematical modeling), from science to social life (IQ scores, credit ratings, performance metrics), and from social life to the most intimate dimensions of personal experience (calorie counting, step tracking, sleep scoring, mood rating).

The quantification imperative, the belief that understanding something means measuring it, and that measuring it means assigning it a number, is one of the deepest legacies of monetary civilization. And it is a legacy that AI both inherits and amplifies.

AI systems can only process what can be quantified. An AI trained on numerical data can identify patterns, make predictions, and generate recommendations. An AI trained on text can process language, but only language that has been digitized, tokenized, and converted into numerical representations. Everything that resists quantification — the quality of a sunset, the weight of a silence, the meaning of a particular facial expression in a particular context, falls outside the system's competence.

This is not a limitation of current AI that future AI will overcome. It is a structural feature of computational arrangements: they operate on formal representations, and formal representations are, by definition, quantitative. The world as experienced by a human being is qualitative — rich with meaning, context, nuance, and the kind of significance that resists numerical capture. The world as processed by a computational infrastructure is quantitative — reduced to patterns in data that can be analyzed, compared, and optimized.

The gap between these two worlds, the qualitative world of human experience and the quantitative world of computational representation, is one of the central tensions of the digital age. And it is a tension that money first created, five thousand years ago, by establishing the principle that unlike things could be compared on a shared numerical scale.

Money taught civilization to count. Science taught it to measure. Computers taught it to compute. AI teaches it to optimize. The trajectory is consistent: each step deepens civilization's commitment to quantification as the primary mode of understanding. And each step makes the unquantifiable — the aspects of human experience that resist numerical capture, less visible, less valued, and less institutionally powerful.

The person who cannot justify their intuition with data is, in a modern institutional setting, at a disadvantage. The employee who cannot demonstrate their value through metrics is vulnerable. The student who cannot prove their understanding through test scores is undercounted. The citizen who cannot articulate their concern in terms that fit a cost-benefit analysis is unheard. In each case, the problem is not that quantification is wrong. It is that quantification is incomplete, and that institutions, having been trained by five millennia of monetary logic and a century of computational logic to trust numbers over narratives, systematically privilege the quantifiable over the qualitative.

Money began this process. AI may complete it. And the question of whether anything important about human experience will survive the completion is one of the most decisive questions this book can raise.

That seam is where much of modern life hides its discomfort.

Money also changes how people imagine the future. Wealth can be stored. Obligation can compound. Risk can be priced. Time itself begins to feel more convertible. A society operating through money can grow more flexible, but it can also become more comparative, more anxious, and more skilled at translating human realities into negotiable units.

The social transformation here is deeper than accounting. Once value can travel in abstract form, obligation no longer remains trapped inside local memory or visible exchange. A debt can persist without friendship. A transaction can proceed without reverence. A hierarchy can stabilize through price rather than lineage alone. Money does not abolish older forms of power. It makes them more mobile.

The reason is that monetization so often arrives as liberation and matures as comparison. The first feeling is often relief. Friction falls. Exchange widens. Dependence on particular persons lessens. But over time, a subtler discipline enters. The person begins to evaluate not only goods, but labor, time, competence, and even self-worth through increasingly commensurable measures. Once enough of life can be translated into units, the imagination itself starts speaking in units.

That does not mean everything becomes market logic in a total sense. It means market logic gains astonishing reach.

And reach changes character. A village may know who is generous, dangerous, noble, cruel, or useful. A monetized system can know what is owed, what is priced, what is payable, and what is profitable. These are not morally identical kinds of knowledge. They coexist, but one gradually becomes easier to operationalize at scale.

That operational ease is one of money's great hidden powers.

The brilliance of money is not only that it moves value. It makes value administrable.

This is what links money so naturally to states, armies, taxation infrastructures, debt regimes, accounting practices, and long-distance trade. Once value can be stabilized in abstract form, it can be recorded, stored, compared, enforced, accumulated, and inherited with a new kind of rigor. The fluidity looks like freedom because it is freeing in many respects. Yet it also creates one of civilization's oldest temptations: to assume that what can be priced is what matters most clearly.

That temptation is not always conscious. It often arrives as habit.

The person who begins by using money to simplify exchange may end by experiencing simplification itself as moral clarity. Numbers can feel cleaner than stories. Price can feel firmer than context. Quantification can feel more trustworthy than memory. Under the right conditions, abstraction begins to masquerade as neutrality.

The point is where the deepest shift occurs.

Money stops looking like a human invention and starts looking like reality.

And once that happens, the next great revolution follows naturally. If value can be abstracted, stabilized, and circulated, then belief can be as well.

The next framework will do for ideas what money did for exchange.

It will make them massively portable.

PART III

The Industrialization of the Human Condition

The psychological consequences of monetary abstraction run deeper than most economic analyses acknowledge. Think about what it means, experientially, to live in a thoroughly monetized society. Every hour of the day can be converted, at least in principle, into a monetary value. The time you spend sleeping has an "opportunity cost", it is time you could have spent earning. The time you spend with your children has an implicit price — it is time you are not billing. The meal you cook at home can be compared, in monetary terms, to the meal you could have purchased. The walk you take for pleasure could, theoretically, have been a productive hour.

This is not how most people consciously experience their days. But it is the background logic of a monetary civilization, and it exerts a subtle, pervasive pressure on how people evaluate their choices. The person who feels guilty for taking an afternoon off is not experiencing a natural emotion. They are experiencing the internalized logic of monetary valuation — the deep, often unconscious belief that time not converted into economic output is time wasted.

Georg Simmel, the German sociologist, explored this dynamic in *The Philosophy of Money* with astonishing subtlety. Money, Simmel argued, does not merely facilitate exchange. It transforms the mental life of the people who use it. It encourages what he called a "calculative" orientation toward reality, a tendency to evaluate experiences, relationships, and even emotions in quantitative terms. The monetized mind does not merely count money. It counts everything. It compares, measures, ranks, and evaluates with a relentlessness that reflects not personal obsession but civilizational training.

Consider the modern concept of "human capital", the idea that a person's education, skills, and experience can be understood as an investment that yields economic returns. This concept, now standard in economics and human resources, would have been deeply strange to most pre-monetary societies. The idea that a human being is, in some meaningful sense, an asset, that their value can be calculated, their potential quantified, their worth measured against their cost, is a product of monetary thinking so thoroughly absorbed that it no longer feels like a metaphor. It feels like a description.

But it is a metaphor. And metaphors shape perception. When we speak of "investing" in relationships, "spending" time, "paying" attention, or "earning" trust, we are not merely using convenient figures of speech. We are experiencing the world through a conceptual framework that money created, a framework in which everything, including the most intimate dimensions of human experience, can be translated into the language of exchange.

The monetization of time is perhaps the most consequential of these translations. Before money, time was experienced primarily as duration — the passage of seasons, the aging of the body, the rhythm of work and rest. Money introduced a new way of experiencing time: as a resource that could be saved, spent, wasted, or invested. The clock and the coin are, in a sense, companion technologies — both quantifying something that was previously experienced as qualitative, both enabling forms of coordination that depend on standardized measurement, and both fundamentally altering the inner experience of the dimension they measure.

When Benjamin Franklin wrote "time is money" in 1748, he was not describing an eternal truth. He was describing a recent invention — a way of thinking about time that had emerged alongside monetary economies and industrial production. But the invention has been so thoroughly absorbed that Franklin's phrase now sounds not like an insight but like an observation. That is the mark of a fully successful revolution: the invented feels discovered, the constructed feels natural, and the metaphor feels like the literal truth.

This part follows the revolutions that accelerated replication, measurement, production, and influence. The world becomes faster, more standardized, more scalable, and more capable of reorganizing human beings around external orders.

By this stage, civilization is no longer merely learning to store, count, and coordinate. It is learning to accelerate itself. That acceleration will reshape truth, labor, time, persuasion, and eventually the emotional architecture of daily life.

CHAPTER 7
The Printing Press: When Belief Became Reproducible

Before print, knowledge traveled more slowly, more unevenly, and under heavier institutional constraints.

Imagine the shock of seeing one text become many, not through months of copying, but through a process that suddenly made repetition mechanical. A claim that once moved through laborious human hands could now multiply with a force that felt almost biological. That is the psychological threshold the printing press opened. Manuscripts existed, ideas circulated, and traditions persisted, but duplication was laborious. Reproduction carried friction.

The printing press changed that by making belief reproducible.

Think about what it meant to produce a book in medieval Europe. A single manuscript Bible required months of skilled labor. A scriptorium might produce a handful of volumes per year. Each copy was precious and expensive, under the authority of whatever institution commissioned it.[18]

Elizabeth Eisenstein, in *The Printing Press as an Agent of Change*, argues that print did not merely speed existing processes. It transformed them, creating new relationships between authors and audiences, new economics of knowledge, and new politics of dissent.[19]

By 1500, roughly fifty years after Gutenberg, an estimated twenty million volumes had been printed in Europe. By 1600, perhaps two hundred million. A literate European in 1400 might encounter a few dozen books in a lifetime. By 1550, hundreds of titles were available in any major city. The world of available knowledge had expanded by orders of magnitude within two generations.

Consider Martin Luther. In October 1517, his Ninety-Five Theses challenged specific Church practices. Within weeks, they had been translated, set in type, and printed in multiple editions across

German-speaking territories. One scholar with an argument had been amplified by the press into a force the most powerful institution in Western Christendom could not suppress. Luther's success was not only theological. It was logistical, he understood that print had changed what was possible in the distribution of belief.

Print also reshaped truth itself. A printed page can feel firmer than speech because it appears less vulnerable to distortion. Mechanically repeated error can feel more authoritative than fragile oral truth. Reproducibility does not merely scale knowledge. It scales confidence. Vernacular literatures flourished. Standard languages solidified around printed norms. Scientific findings could be published, compared, and built upon across Europe within months.

This does not mean truth automatically spread. It means ideas acquired a new industrial property: scale.

Scale changes the psychology of conviction. A text copied by hand remains precious and limited. A text mechanically reproduced becomes socially aggressive. It can travel fast, multiply, compete, saturate, provoke, and standardize. The printing press turns communication from an artisanal process into a replicating force.

What it prints can be scripture, science, propaganda, heresy, reform, myth, bureaucracy, or instruction. The medium does not guarantee wisdom. It guarantees velocity.

There is another dimension of the print revolution that is rarely discussed but that is directly relevant to the later chapters of this book: the transformation of intellectual authority from persons to texts.

In a manuscript culture, intellectual authority is closely tied to personal identity and institutional affiliation. A claim carries weight because of who makes it, which scholar, from which university, with what credentials, under what patronage. The scarcity of texts means that relatively few voices participate in public intellectual life, and those who do are usually vetted by institutional gatekeepers.

Print loosened this connection between authority and identity. A pamphlet could be published anonymously. A book could circulate without its author being known. Ideas could spread on their own merits, or on their own inflammatory appeal — without the anchor of personal reputation. This was liberating in many respects: it allowed dissidents to challenge power without risking immediate personal consequences. It allowed ideas to be evaluated (at least in principle) on their merits rather than their provenance. It democratized intellectual life by lowering the barriers to entry.

But it also created a new problem: the problem of authentication. In a manuscript culture, the question "who wrote this?" has a relatively clear answer — the scribe, the author, the institution. In a print culture, the question becomes more complicated. Forgeries proliferate. Anonymous pamphlets circulate. Attributed texts may be altered or fabricated. The reader must develop new skills of critical evaluation — skills that go beyond simply trusting the source and extend to evaluating the content itself.

This problem, the gap between the proliferation of claims and the capacity to evaluate them, is the central epistemological challenge of every media revolution that follows. It appears with newspapers (which are often accused of bias, fabrication, and sensationalism), with radio (which can broadcast propaganda as effectively as truth), with television (which privileges appearance over substance), with the internet (which eliminates editorial gatekeeping entirely), and with AI (which can generate unlimited quantities of fluent, persuasive text with no commitment to accuracy).

Print was the first technology to create this gap. And the strategies that print culture developed to address it, peer review, editorial standards, citation practices, institutional reputation, libel law, are the ancestors of every subsequent attempt to manage the problem of trust in a world of abundant, unverified information.

The history of print is therefore not merely the history of a technology. It is the history of a recurring civilizational challenge: how to maintain shared truth in a world where the tools for producing and distributing claims have outrun the tools for evaluating them. Every chapter from this point forward in the book confronts a version of this challenge — each time more acute, each time faster, each time harder to resolve.

And so print matters so deeply to this book's argument. It does not merely increase access to information. It alters the structure of authority.

Once texts can proliferate beyond old bottlenecks, monopolies on truth begin weakening. Clergy, courts, monarchies, and inherited interpreters no longer control meaning as completely as before. Literacy gains new importance. Standard languages become more possible. Public argument widens.

There is one more concrete scene from the print revolution that illuminates its transformative power: the experience of reading in solitude.

Before print, reading was, for most people, a social activity. Texts were read aloud, in churches, in courts, in gathering places, in the presence of others. The experience of encountering ideas was communal. Meaning was negotiated collectively. The reader's response was shaped by the reactions of those around them.

Print made solitary reading possible on a mass scale. For the first time, millions of individuals could encounter ideas alone, in private, without the mediation of an institutional context or the influence of a surrounding audience. The reader could think their own thoughts about what they read. They could disagree without being overheard. They could be persuaded without being observed. They could develop interpretations that deviated from the orthodox without risking immediate social sanction.

This privatization of reading was one of the most consequential psychological shifts in Western history. It created the conditions for individual conscience, the experience of holding beliefs that were arrived at privately, through personal engagement with text, rather than collectively, through communal participation in oral culture. The Protestant Reformation's emphasis on individual reading of scripture, on the believer's direct, personal, unmediated encounter with the word of God, was only possible in a culture where solitary reading had become a widespread practice.

The privatization of reading also created the conditions for the modern concept of the self, the idea that each individual possesses an interior life that is distinct from their social performance, that has its own thoughts, its own responses, its own conclusions. This concept, which modern people take entirely for granted, is historically specific. It depends on the existence of a practice — solitary reading, that allows the individual to develop an inner life that is not continuously shaped by social interaction.

The digital age has, in many respects, reversed this privatization. Reading online is increasingly social — shared, commented upon, liked, retweeted, algorithmically recommended. The solitary encounter with a text, the experience of reading a book alone, in silence, forming one's own thoughts without the real-time influence of others' reactions, is becoming rarer. And with its decline, the conditions that produced individual conscience, private thought, and the modern concept of the self may be quietly eroding.

This creates immense intellectual energy. It also creates conflict.

There is one more dimension of the print revolution that connects it to the present with particular force: the invention of intellectual property.

Before print, the concept of "owning" an idea was nearly meaningless in practical terms. A storyteller who told a tale did not own the tale. A scholar who wrote a treatise might be honored for it, but the treatise, once copied by hand, belonged as much to the copyist as to the author. The scarcity of copies, combined with the labor required to produce them, made the question of intellectual ownership largely moot.

Print changed this by making reproduction cheap and copies numerous. If a publisher could produce a thousand copies of a book in the time it took a scribe to produce one, then the economic value of the text, as distinct from the economic value of any individual copy, became significant. And if the text had economic value, then the question of who owned that value became urgent.

The result was the gradual development of copyright law — a legal framework for assigning ownership to symbolic creations. The Statute of Anne, enacted in England in 1710, is generally considered the first modern copyright law. It established that authors, rather than publishers or the Crown, held the primary right to control the reproduction of their works. This was a revolutionary principle: the idea that a pattern of symbols, not the paper it was printed on, not the ink that formed the letters, but the arrangement of ideas itself — could be owned.

This principle has shaped civilization ever since. The modern economy of ideas, the economy that produces books, music, software, films, pharmaceutical formulas, engineering designs, and every other form of creative and intellectual output, rests on the legal fiction that intangible patterns can be owned. And the conflicts surrounding that fiction — from music piracy to software patents to the training data used by AI systems, are among the most consequential legal and ethical disputes of the contemporary world.

Print invented the problem. Every subsequent technology has intensified it. The internet made copying essentially free. AI made generation nearly costless. The question of who owns an idea, and what "ownership" even means when ideas can be reproduced infinitely at zero marginal cost — is a question that print first posed and that civilization has never fully answered.

Print also fundamentally altered the economics of intellectual life. Before print, scholars depended on patronage, the support of wealthy individuals, religious institutions, or royal courts. Print created the possibility of an independent intellectual market: a world in which ideas could be sold directly to readers, and in which the commercial success of a book could provide an income independent of any patron's favor.

This was liberating in many respects. It freed intellectuals from the constraints of patronage and allowed them to address audiences of their own choosing. It created the possibility of the professional writer, the independent journalist, and the public intellectual — figures whose livelihood depended on the quality and popularity of their work rather than on the goodwill of powerful sponsors.

But it also subjected intellectual life to market pressures. A book that sells well is not necessarily a book that is true, wise, or beneficial. A pamphlet that inflames is often more commercially successful than one that carefully qualifies. The market for ideas, like all markets, tends to reward what people want to buy rather than what they need to know. And the tension between intellectual quality and commercial viability — between what is good and what sells — has been a defining feature of intellectual life ever since.

This tension recurs, in amplified form, in every subsequent media revolution. The internet created an attention economy in which the most-clicked content is the most-visible content, regardless of its quality. Social media created an engagement economy in which the most-shared content is the most-influential content, regardless of its accuracy. AI is creating a generation economy in which the most content can be produced at the lowest cost, regardless of its depth or originality.

In each case, the economic logic of the medium shapes the intellectual environment in ways that are not determined by the intentions of any individual participant. No one set out to create a media ecosystem that rewards sensationalism over substance. But the economic structure of attention-based media produces that result as reliably as gravity produces falling. Print was the first medium to demonstrate this dynamic. Its descendants have only intensified it.

The printing press democratizes access to claims, not necessarily to discernment. Once belief becomes reproducible, conviction becomes more contagious. Dissent becomes more scalable. Orthodoxy becomes harder to preserve. Persuasion becomes semi-industrial.

In this way, print is one of history's first great information shocks. It teaches civilization a lesson that will repeat in later media revolutions: when expression expands, control structures do not simply vanish. They adapt, react, censor, compete, and fracture.

There is a dimension of the print revolution that anticipates the AI revolution with uncanny precision: the question of what happens to authority when the cost of producing convincing text drops to near zero.

In a manuscript culture, the production of text is expensive, laborious, and institutionally controlled. This means that the volume of text in circulation is limited, and that most of the text that does circulate has passed through some form of institutional filtering — whether by a monastery, a patron, a university, or a royal court. The filtering is imperfect and biased, but it provides a rough quality signal: a text that has been copied, preserved, and circulated by a serious institution is more likely to be of value than a random set of marks on a wall.

Print disrupted this quality signal by making text production cheap. The pamphleteer, the broadside writer, the penny-press journalist, all could now produce and distribute text at a fraction of the cost that manuscript culture required. The volume of available text exploded. And the institutional filtering that had previously served as a rough quality signal was overwhelmed.

The result was a crisis of authority, a period in which the old mechanisms for determining what was trustworthy had been disrupted, and new mechanisms had not yet been established. That crisis produced, among other things, the Reformation, the wars of religion, the development of scientific method as a new foundation for knowledge claims, and the eventual emergence of the modern press as an institution with (variable) standards of accuracy and accountability.

AI is producing an analogous crisis. The cost of generating convincing text, images, audio, and video has, with generative AI, dropped to near zero. The volume of content that can be produced by a single individual or a small team has expanded by orders of magnitude. The institutional filtering mechanisms that developed over centuries to manage the print revolution — editorial standards, peer review, journalistic ethics, copyright law, are being overwhelmed by a technology that produces content at a pace and scale they were never designed to handle.

The parallels with the early print era are striking. The pamphlet wars of the sixteenth century, in which competing factions produced torrents of persuasive, inflammatory, and often fabricated material, are structurally identical to the information wars of the social media age — and AI is about to intensify those wars by orders of magnitude. The crisis of authority that print produced, the breakdown of shared epistemological foundations, the fragmentation of public discourse into competing realities, is being replicated, at far greater speed and scale, by AI.

If history is any guide, the resolution will take decades. The print revolution's crisis of authority was not resolved in a generation. It required the development of new institutions (the scientific society, the modern university, the professional press), new methods (empirical investigation, peer review, editorial standards), and new social norms (the expectation of citation, the ideal of objectivity, the concept of journalistic integrity). Each of these developments took time, time to be invented, time to be institutionalized, and time to be absorbed into the culture's expectations.

AI's crisis of authority will likely require a similar period of institutional innovation. The old mechanisms — editorial gatekeeping, institutional reputation, expert credentialing, will not disappear, but they will prove insufficient for a world in which convincing content can be generated by anyone, at any scale, at near-zero cost. New mechanisms — perhaps involving cryptographic verification, AI-powered fact-checking, digital provenance tracking, or institutional forms that do not yet exist, will need to be developed.

And in the interim, civilization will have to navigate a period of profound epistemological uncertainty — a period in which the question "how do I know this is true?" has no reliable answer, and in which the ability to distinguish signal from noise, truth from fabrication, and expertise from performance becomes one of the most valuable and most endangered human capabilities.

Print created the first version of this crisis. AI is creating the latest. The pattern is the same. The stakes are higher.

Print multiplies realities.

This is not a metaphorical flourish. It is a social fact. A world in which one text dominates is structurally different from a world in which rival texts can circulate broadly and repeatedly. People can now belong to communities of interpretation wider than their local surroundings. Reform becomes imaginable. Counter-interpretation becomes thinkable. The page can now destabilize the throne and the altar at a distance.

At the same time, print creates standardization. A printed text holds shape. A language can be regularized. An argument can be reproduced with fidelity. The press introduces both pluralism and discipline. It opens discourse while hardening form.

That dual nature makes it historically explosive. The same medium that liberates thought can intensify dogma. The same process that spreads scholarship can spread fanaticism. The same tool that educates can inflame.

The lesson is familiar now. Media revolutions do not merely transmit content. They reconfigure what societies can collectively sustain as believable.

One reason print mattered so much is that it altered the economics of persistence. In manuscript culture, a text could matter deeply while still remaining materially scarce. Print loosened that scarcity. A pamphlet could now move through a population with a speed previously reserved for rumor, but with the durability and apparent authority of written form. This combination was historically combustible. Spoken controversy could now be stabilized in print, re-read, cited, copied, attacked, and defended.

That changed the temperature of public conflict.

Belief no longer traveled only through sermons, decrees, and local transmission. It could circulate in portable, repeatable, semi-standardized form. The reader could encounter doctrine without direct institutional mediation. The dissenter could acquire allies at a distance. The local grievance could become a distributed argument.

This is why the printing press belongs in the lineage of normalization rather than merely in the history of literacy. It teaches populations to live amid multiplied claims. It widens access to thought while reducing the old frictions that once kept certain kinds of conviction contained.

The page becomes both liberator and accelerant.

Print also reshapes the psychology of truth by stabilizing text itself. A printed page can feel firmer than speech because it appears less vulnerable to distortion. This gives print extraordinary authority even when what is printed is false or inflammatory. Mechanically repeated error can feel more objective than fragile oral truth.

That is one of print's darkest lessons.

Reproducibility does not merely scale knowledge. It scales confidence.

As printed culture matures, societies begin adjusting to a world in which rival claims no longer disappear simply because old institutions reject them. Counterworlds can now survive in public. Standardization and fragmentation grow together. That apparent contradiction is one of modernity's recurring signatures.

Print pushed Europe and, in time, the wider world into a new relation with authority. Once sacred and political claims could be duplicated, contested, and redistributed at speed, a deeper question emerged: if truth can no longer rely solely on inherited institutions, what new basis will it have?

The answer would not come all at once, but one of its most decisive forms would emerge through method.

Science.

The full scope of what print unleashed is difficult to appreciate from inside a world that takes mass literacy for granted. Before print, a literate European who wished to read Plato needed access to a hand-copied manuscript housed in a monastery or wealthy patron's collection. Two copies of the same work might differ significantly due to generations of copyist errors. After print, hundreds of copies were identical. Errors could be corrected in subsequent editions. Scholars in different cities could discuss the same text with confidence they were reading the same words.

But print also revealed a truth that recurs with every media revolution: expanding access to information does not automatically improve discourse. The same presses that printed Luther's Bible printed virulent anti-Semitic pamphlets. The same technology that enabled the scientific revolution enabled an explosion of astrological quackery and conspiracy. The pamphlet wars of the sixteenth century were, in many respects, the Twitter threads of their day — fast, cheap, emotionally charged, and resistant to verification.

When a medium democratizes expression, it democratizes all expression, the wise and the foolish, the true and the false. The press does not distinguish between scripture and sedition. It reproduces both with equal fidelity. The political consequences were seismic: the English Civil War, the French Revolution, and the American Revolution were all, in part, media events shaped by mass-circulated printed arguments.

Print also created the modern concept of the public, a body of citizens sharing access to the same information and therefore capable of collective deliberation. The philosopher Habermas described this as the emergence of the public sphere — a space of discourse between household and state, made possible by the circulation of printed materials in coffeehouses, salons, and the mass press. Whatever one thinks of Habermas's account, the core insight holds: print created a form of collective consciousness in which people who would never meet could participate in shared conversation about truth, justice, and governance. That conversation continues. Its contemporary form, conducted through digital platforms, carries the same tensions and the same dangers that print first introduced five centuries ago.

Print pushed civilization into a new relationship with authority. If truth could no longer rely solely on inherited institutions, what new basis would it have?

One answer would emerge through method rather than revelation.

Science.

CHAPTER 8

Science: When Mystery Lost Authority

Every civilization has explanations. Science transformed not only the quality of those explanations, but the criteria by which explanations earned the right to govern belief.

The point is what makes the scientific revolution so important. It did not merely discover more about the world. It challenged the social authority of inherited certainty.

Before method acquires prestige, truth can remain anchored in ancestry, scripture, custom, interpretation, or institutional standing. Once method becomes central, a different standard begins asserting itself: observation, experiment, replication, measurement, prediction.

The result is not simply an intellectual shift. It is a political one.

Who gets to define reality after science takes hold? Not only the priest, the elder, the monarch, or the commentator, but the investigator with procedures that can survive scrutiny.

Science dethrones by method.

Consider the world of Galileo. When he turned his telescope toward Jupiter in January 1610 and observed four moons orbiting the planet, he was producing evidence that contradicted the Aristotelian-Ptolemaic model endorsed by the Catholic Church. The conflict was not merely between reason and superstition but between two systems of legitimacy, one based on tradition and scripture, the other on reproducible observation.

The Royal Society of London, established in 1660, formalized the shift with its motto: *Nullius in verba*, take nobody's word for it. Authority would rest on demonstration and communal scrutiny.

Consider medicine. William Harvey demonstrated blood circulation in 1628, overturning fourteen centuries of Galenic authority. Edward

Jenner developed vaccination in 1796. Germ theory replaced miasma theory. Surgery became survivable with antisepsis. Public health interventions — clean water, sewage systems, saved millions. Each advance depended on shared standards, peer review, and the willingness to revise belief when evidence demanded it.

Thomas Kuhn, in *The Structure of Scientific Revolutions*, showed that science advances through paradigm shifts that restructure what counts as legitimate inquiry.[43] Each paradigm, once established, becomes the invisible background of normal research — demonstrating the pattern of this book within the most successful knowledge system humanity has produced.

Science is not just a body of knowledge. It is a civilizational permission structure. It teaches that the world can be known systematically and therefore altered systematically. Mystery loses authority. Intervention gains legitimacy.

This yields enormous human gains. Medicine improves. Astronomy corrects inherited cosmologies. Physics enlarges predictive power. Technology grows from disciplined understanding. Disease can be studied rather than merely endured or interpreted morally. Matter becomes less mysterious and more usable.

Yet every expansion of usable knowledge also expands the moral confidence of intervention.

Once the world becomes measurable, it becomes harder to leave alone.

That is the deeper continuity between science and later technological revolutions. Science is not just a body of knowledge. It is a civilizational permission structure. It teaches that the world can be known systematically and therefore altered systematically.

Mystery loses authority. Intervention gains legitimacy.

Science belongs in this logic not as an exception, but as one of its most decisive accelerants. Scientific method feels almost self-evidently good in the modern world because its successes are so overwhelming. But that success also normalizes a habit of mind: the habit of treating reality as open to analysis, manipulation, and optimization.

The world becomes less enchanted and more actionable.

To understand the force of that change, it helps to see what science displaces. In inherited machinerys of certainty, many explanations derive authority from sanctity, lineage, revelation, or long-held custom. Science does not merely offer better answers in some domains. It changes what counts as a legitimate answer. The burden of persuasion begins shifting from inherited standing to demonstrable procedure.

The point is an extraordinary transfer of authority.

It does not happen instantly or cleanly. Old organizations remain powerful. New methods are resisted, absorbed, institutionalized, and sometimes mythologized in turn. But over time, the cultural prestige of method grows. To know becomes less a matter of rightful inheritance and more a matter of disciplined demonstration.

There is one final scene that captures the scientific revolution's psychological impact. In 1859, Charles Darwin published *On the Origin of Species*. The book's argument, that human beings were not specially created but had evolved, through natural selection, from the same processes that produced every other living thing — was, for many readers, the most psychologically disturbing claim in the history of science.

The disturbance was not merely intellectual. It was existential. If human beings were products of the same blind, purposeless processes that produced worms and bacteria, then the special status that religion, philosophy, and common sense had assigned to the human species was an illusion. Humans were not the purpose of creation. They were an accident of it, one branch on a vast evolutionary tree, distinguished from other branches by degree rather than kind.

This was, for many people, unbearable. And the resistance it provoked, resistance that continues, in various forms, to the present day, reveals something important about the relationship between knowledge and psychological comfort. The theory of evolution is, by the standards of scientific evidence, one of the best-supported theories in all of science. And yet large portions of the population in many countries reject it, not because the evidence is weak, but because the implications are intolerable.

This gap between evidence and acceptance is a recurring feature of the scientific revolution's psychological legacy. Science repeatedly produces findings that are well-supported by evidence and deeply disturbing to human self-image. The Earth is not the center of the universe. Human beings are not specially created. Consciousness may be a product of brain activity rather than an immaterial soul. Free will may be more limited than common sense suggests. These findings are, by scientific standards, well-established. By psychological standards, they are almost intolerable.

The result is a civilization that simultaneously depends on science for its material prosperity and resists science when its findings threaten psychological comfort. This tension, between the power of scientific knowledge and the fragility of human self-regard — runs through every chapter of this book from this point forward. AI threatens the special status of human intelligence. Synthetic intimacy threatens the special status of human connection. Predictive networks threaten the special status of human choice. Singularity threatens the special status of the human itself.

In each case, the pattern is the same: a technology reveals that something humans believed was unique and irreplaceable may be, in fact, replicable, approximate, or contingent. And in each case, the psychological response follows the pattern that Darwin's theory first provoked: not calm evaluation of the evidence, but existential resistance to the implications.

This yields not only knowledge, but confidence. The more the world yields to method, the more human beings begin to trust intervention. The body can be treated. The sky can be calculated. Motion can be modeled. Matter can be broken down, recombined, and harnessed. Scientific success trains the civilization-scale intuition that reality is not only meaningful, but workable.

That intuition is one of modernity's deepest inheritances.

It is also one of its most dangerous seductions when severed from restraint.

A measurable world invites optimization. An optimizable world invites administration. An administrable world invites the extension of systems into domains once protected by mystery, custom, or moral hesitation. Science itself does not dictate how power uses its discoveries, but it enlarges the scope of what power can plausibly attempt.

This is why the chapter cannot be written as a simple celebration or a simple lament. Science frees humanity from immense ignorance and suffering. It also weakens older barriers to intervention. It can make the world clearer while making human beings bolder in ways that outpace wisdom.

The modern world often mistakes scientific legitimacy for moral sufficiency. That confusion matters. A thing can be possible to measure and possible to manipulate without being obviously worthy of manipulation. Yet successful method quietly teaches civilizations to trust capability.

There is a further dimension of the scientific revolution that connects directly to the digital age: the relationship between specialization and public understanding.

As science became more successful, it also became more specialized. The natural philosopher of the seventeenth century — a generalist who might contribute to optics, mathematics, theology, and politics in the same lifetime, gave way to the specialist of the nineteenth and twentieth centuries: the organic chemist, the particle physicist, the molecular biologist, the econometrician. Each field developed its own vocabulary, its own methods, its own standards of evidence, and its own institutional infrastructure. The result was an explosion of knowledge that was also, paradoxically, an explosion of ignorance.

Let me explain what I mean by that paradox. In 1700, an educated person could, in principle, understand the major findings of every scientific discipline that existed. The total body of scientific knowledge was small enough, and the methods simple enough, that a generalist could achieve reasonable competency across multiple fields. By 1900, this was already impossible. By 2000, it was laughable. The total body of published scientific research is now so vast that no human being can read even a fraction of the papers published in a single discipline in a single year.

This means that the civilization whose most powerful institutions — medicine, engineering, agriculture, military, commerce, governance — depend on scientific knowledge is also a civilization in which no individual, and no group of individuals, can understand the full body of knowledge on which those institutions depend. The doctor who prescribes a medication does not know the physics of drug synthesis. The engineer who designs a bridge does not know the biology of the materials scientists who developed the steel alloy. The voter who evaluates climate policy does not know atmospheric chemistry.

Each of these individuals trusts, must trust, the specialists in other fields. And that trust is, for the most part, well-placed. The system of scientific specialization, peer review, and institutional reputation provides a reasonably reliable mechanism for producing and validating knowledge, even when no single person can verify all of it.

But this structural dependency on trust is also a vulnerability. When trust in scientific institutions erodes — as it has, in recent decades, around issues like vaccination, climate change, nutrition science, and pandemic response — the consequences are not merely intellectual. They are civilizational. A population that does not trust its scientific institutions is a population that cannot respond coherently to scientific problems, including problems that threaten its survival.

The irony is acute. Science, which was born as a challenge to the authority of inherited tradition and institutional power, has itself become an institutional authority that demands trust. The Royal Society's motto, "take nobody's word for it" — was a revolutionary declaration against deference. Three centuries later, science asks the public to do precisely what it once refused to do: take the word of authorized experts on matters too complex for individual verification.

This is not hypocrisy. It is the inevitable consequence of knowledge growing beyond the comprehension of any individual mind. But it creates a tension that is directly relevant to the AI revolution: if the public must already trust scientific claims it cannot verify, how much more must it trust algorithmic decisions whose logic it cannot even inspect?

The black box of AI is not a new problem. It is an intensification of a problem that science created: the gap between the power of expert systems and the public's ability to evaluate them. AI merely makes the gap wider, faster, and more consequential.

There is one final dimension of the scientific revolution that bears directly on the present: science's complicated relationship with uncertainty.

Science is, at its core, a method for managing uncertainty. It does not produce absolute truths. It produces provisional conclusions — claims that are supported by current evidence and subject to revision when new evidence arrives. This is one of science's greatest strengths: its willingness to change its mind, to admit error, and to revise even its most cherished theories in the face of contradictory data.

But this strength is also, in the public sphere, a vulnerability. A population that expects certainty, and that has been trained by generations of technological success to associate science with reliable answers, can be deeply unsettled by scientific uncertainty. When scientists disagree, when recommendations change, when provisional conclusions are revised, the public response is often not appreciation for the honest management of uncertainty but frustration, confusion, and loss of trust.

This dynamic was visible during the COVID-19 pandemic, when public health recommendations changed repeatedly as new evidence emerged. Mask guidance shifted. Transmission understanding evolved. Vaccine recommendations were updated. Each change was, from a scientific perspective, a rational response to new data. From a public perspective, it often felt like inconsistency, incompetence, or even deception.

The gap between how science manages uncertainty and how the public experiences uncertainty is one of the central challenges of modern governance. And it is a gap that AI may widen rather than narrow. AI architectures produce outputs with apparent confidence, clean numbers, definitive recommendations, fluent prose. They do not naturally communicate uncertainty, nuance, or the limitations of their training data. A user who asks an AI system a question receives an answer, not a discussion of the evidence, not a range of possibilities, not a careful qualification of what is known and what is not. The answer may be hedged with disclaimers, but the format, the confident, fluent, authoritative presentation — communicates certainty regardless of the disclaimers' content.

If the public already struggles with scientific uncertainty, AI may make the struggle worse by producing a generation accustomed to receiving confident answers to every question, answers that feel authoritative, that arrive instantly, and that do not require the uncomfortable work of sitting with ambiguity, evaluating competing claims, or accepting that some questions do not have clear answers.

Science's greatest lesson is that uncertainty is not a failure of knowledge. It is a feature of honest inquiry. The question is whether a civilization increasingly shaped by systems that produce the appearance of certainty will retain the capacity to value — and to tolerate — the reality of uncertainty.

There is also a psychological cost to this transfer. Mystery does not vanish merely because explanations improve. It loses status. The unknown stops appearing sacred by default and begins appearing temporary, technical, or solvable. That reclassification changes the emotional weather of civilization. Wonder remains possible, but it competes with a new expectation: that enough knowledge will eventually yield control.

There is a further consequence of the scientific revolution that connects directly to the AI era: the creation of what might be called "expertise dependence."

In a pre-scientific society, the knowledge required for daily decision-making is, for the most part, accessible to the individual. A farmer understands the soil, the weather, and the crops. A craftsman understands the materials and techniques of their trade. A healer understands the herbs and practices of their tradition. The knowledge may be incomplete, but it is personally possessed and personally verifiable. The individual is, in a meaningful sense, the author of their own practical knowledge.

The scientific revolution changed this by demonstrating that expertise, specialized, methodically acquired, institutionally validated knowledge, produces better outcomes than common sense in domain after domain. Scientific agriculture outperforms traditional farming. Scientific medicine outperforms folk healing. Scientific engineering outperforms intuitive construction. In each domain, the advantage of specialized expertise is so overwhelming that relying on personal judgment alone becomes, over time, not merely suboptimal but irresponsible.

The result is a civilization in which the individual is expected to defer to expert judgment in an ever-expanding range of domains: medicine, nutrition, child-rearing, financial planning, legal compliance, technological adoption, environmental behavior, psychological well-being. The modern citizen is not merely encouraged to consult experts. They are, in many cases, legally required to do so, to consult a doctor before taking certain medications, a lawyer before signing certain contracts, an accountant before filing certain tax returns.

This expertise dependence is, in many respects, a good thing. Expert knowledge genuinely produces better outcomes in most domains. The alternative, every individual making medical, legal, and financial decisions based solely on personal judgment, would be catastrophic. The case for expertise is strong and well-supported.

But expertise dependence also carries costs that are rarely acknowledged. The most important is the gradual atrophy of what might be called practical wisdom — the capacity of individuals to make competent decisions in their own lives based on their own experience and judgment. A person who has been trained, from childhood, to defer to experts in every domain may eventually lose confidence in their own capacity for judgment. They may become, in a psychological sense, dependent, unable to make a significant decision without first consulting someone who claims specialized knowledge.

This dynamic is directly relevant to the AI revolution. AI can be understood as the ultimate expert, an order that claims competence across every domain, that offers advice on every question, and that is available at all times for consultation. The person who has already been trained by the scientific revolution to defer to expertise finds in AI the perfect object of deference: an omnipresent, infinitely patient, apparently knowledgeable advisor that requires no appointment, charges no fee, and never expresses frustration with the question.

The risk is not that AI will give bad advice (though it sometimes will). The risk is that pervasive AI consultation will accelerate the atrophy of individual judgment that expertise dependence has already begun. A person who checks an AI system before making every decision, what to cook, how to respond to an email, whether to take an umbrella — is a person who is outsourcing judgment at a rate and scale that no previous generation of experts could have supported. The convenience is real. The cost — in terms of the individual's confidence in their own capacity to navigate the world, may be equally real, though harder to measure.

Science created the template for expertise dependence. AI may complete it.

This expectation is not irrational. It has been rewarded repeatedly. But repeated reward can become cultural overconfidence.

Once enough domains submit to method, the human mind begins treating intervention not as a serious threshold, but as the normal answer to difficulty. The disease should be treated. The crop improved. The machine optimized. The behavior studied. The inefficiency removed. The threshold crossed. Over time, method trains not only inquiry, but impatience with limits.

That impatience will matter later in this book when optimization moves into attention, intimacy, cognition, and personhood itself.

For now, the crucial point is simpler. Science did not merely produce better beliefs. It changed how belief earns authority. That is one of the great civilizational rewrites.

The social reorganization that science demanded was as important as its intellectual achievements. Science required new institutions, universities reorganized around empirical inquiry rather than textual commentary, learned societies that facilitated the exchange of experimental results, journals that published findings for communal scrutiny, and eventually the vast apparatus of modern research: laboratories, funding agencies, peer review structures, international conferences, and standardized citation practices.

Each of these institutions embodied a specific idea about how knowledge should be produced and validated: through collective, transparent, methodical investigation rather than individual revelation or inherited authority. The shift was neither total nor sudden — religious institutions continued to sponsor scientific work, individual genius continued to matter enormously, and the boundaries between science and other ways of knowing remained contested for centuries (and remain contested today). But the direction was unmistakable: authority was migrating from persons to procedures.

This migration has consequences that extend far beyond the laboratory. When a civilization adopts method as its primary criterion for truth, it does not merely change how knowledge is produced. It changes what counts as knowledge. Claims that cannot be tested, moral claims, aesthetic judgments, spiritual experiences, philosophical arguments, do not disappear, but they lose a certain kind of institutional prestige. They become "subjective" in a world that increasingly privileges the "objective." They become "opinion" in a world that increasingly demands "evidence."

This hierarchy, objective over subjective, evidence over opinion, data over intuition — is so deeply embedded in modern culture that it feels self-evident. But it is not self-evident. It is a product of the scientific revolution, and it carries costs alongside its enormous benefits. The cost is a subtle but pervasive devaluation of forms of knowledge that do not submit to quantification or experimental testing. Wisdom, tradition, intuition, aesthetic sensitivity, moral discernment, emotional intelligence — all of these are real forms of knowledge, refined over millennia of human experience. But in a scientifically organized civilization, they occupy a subordinate position. They are "soft" where data is "hard." They are "anecdotal" where statistics are "rigorous." They are "personal" where method is "universal."

This hierarchy matters because it shapes how institutions make decisions. A hospital that evaluates doctors primarily by measurable outcomes may miss the doctor whose greatest skill is in listening, reassuring, and earning the trust of frightened patients. A school that evaluates teachers primarily by test scores may miss the teacher whose greatest impact is on students' character, curiosity, and love of learning. A corporation that evaluates employees primarily by quantified performance metrics may miss the person whose greatest contribution is to team morale, institutional memory, and the kind of judgment that cannot be captured in a dashboard.

The scientific revolution gave civilization an extraordinarily powerful tool for understanding the measurable world. It also installed a cognitive bias — a tendency to treat the measurable as the real and the unmeasurable as the questionable. That bias has served humanity well in many domains. But it has also created blind spots — areas of human experience and value that fall outside the scope of scientific method and are therefore treated as less real, less important, or less worthy of institutional attention.

This is relevant to the later chapters of this book because artificial intelligence, the most powerful product of the scientific-technical tradition, inherits and amplifies this bias. AI mechanisms can only process what can be formalized, quantified, and encoded. Everything that resists formalization — the tacit, the intuitive, the contextual, the emotionally complex, falls outside the framework's competence. And as AI assumes an ever-larger role in institutional decision-making, the bias toward the measurable will intensify, potentially marginalizing the very forms of human judgment that are most needed in a world of increasing complexity.

Once knowledge becomes usable at scale, the next revolution follows with brutal force.

Industry.

CHAPTER 9

Industry: When the Clock Entered the Soul

There are few images more emblematic of modernity than the factory clock.

That is not accidental.

The industrial revolution did not merely produce machines. It reorganized human rhythm around machine-compatible time.

Imagine the sound of a bell determining the moral shape of a day. Not the bell of a church alone, carrying ritual and community, but the bell of a factory shift, the bell that turns lateness into fault, delay into inefficiency, and the body into a timed instrument. That is the atmosphere into which industrial modernity trains people.

Before industrialization, many people still lived under temporal patterns shaped by seasons, weather, custom, and localized demands. Work could be intense and exhausting, but it was often less uniformly synchronized to external precision. Agricultural labor had its own harsh compulsions, but industrial labor introduced something different: a dense, repeatable, externally measured discipline that sought not only effort but regularity.

Industry changes time from condition into command.

The clock becomes sovereign.

This matters because rhythm is not just logistics. Rhythm is a mode of being. It shapes how a person experiences fatigue, expectation, productivity, waiting, obligation, and self-worth. When industrial systems spread, they do not merely alter workplaces. They teach entire societies to think of time as segmented, monetizable, and morally charged.

Time wasted becomes guilt.

Delay becomes inefficiency.

There is one more dimension of the industrial revolution that connects it directly to the digital transformations described in later chapters: the factory's role as a training system for institutional obedience.

The factory did not merely produce goods. It produced a particular kind of person, a person habituated to working under surveillance, following instructions issued by others, performing tasks defined by the needs of a system rather than by personal judgment, and accepting that their time belonged to their employer during designated hours. These are not natural human tendencies. They are learned behaviors, cultivated through years of institutional training that begins in school and is reinforced in the workplace.

Michel Foucault, in *Discipline and Punish*, described this process with unflinching clarity. The modern disciplinary system, Foucault argued, does not primarily operate through spectacular punishment (the public execution, the torture chamber). It operates through routine — through timetables, inspections, standardized procedures, hierarchical observation, and the internalization of behavioral norms that make external coercion increasingly unnecessary. The ideal subject of disciplinary power is not someone who obeys because they fear punishment. It is someone who obeys because obedience has become automatic, because the rules have been internalized so thoroughly that they no longer feel like rules. They feel like the natural structure of the day.

This analysis applies directly to the digital systems described in Parts IV and V. The smartphone does not coerce. It trains. It teaches habits of attention, response, and availability that become automatic through repetition. Social media does not force participation. It creates an environment in which participation feels natural and non-participation feels abnormal. AI does not compel adoption. It makes itself useful enough that not using it begins to feel like a competitive disadvantage.

In each case, the mechanism is identical to the one Foucault described in the factory: power operates not through coercion but through the normalization of behavior. The disciplinary subject does not experience discipline. They experience routine. They experience productivity. They experience the ordinary structure of a normal day. And that experience of normality is itself the product of a machinery that has been designed, whether by factory owners, platform architects, or algorithm engineers — to produce a specific kind of behavior.

The factory's deepest legacy is not the goods it produced or the economic system it enabled. It is the psychological template it established: the expectation that human behavior should be systematic, measurable, optimizable, and aligned with the needs of a larger order. Every subsequent technology of management, from Taylorism to Six Sigma to the quantified self to algorithmic performance monitoring — is a refinement of the factory's original insight: that human beings can be organized as components of a productive network, and that the most efficient organization is one in which the humans no longer experience the organization as externally imposed.

That insight is, in a different vocabulary, the central thesis of this book.

Output becomes virtue.

The machine does not only enter production. It enters aspiration.

Here is one of industrialization's deepest psychological victories. It persuades societies to treat measurable productivity as a major index of human value. The disciplined worker becomes a social model. Regularity becomes moral. Timekeeping becomes identity.

There is one more scene from the industrial era that carries a particular resonance for the digital age: the story of the Luddites.

The Luddites, textile workers in early nineteenth-century England who destroyed mechanized looms in protest against the displacement of their labor and skills, have been thoroughly misremembered by history.

They are invoked as a synonym for irrational technophobia: to call someone a "Luddite" is to accuse them of foolishly resisting the inevitable march of progress.

But the actual Luddites were neither foolish nor irrational. They were skilled craftsmen — people who had spent years mastering the art of hand-weaving, watching their skills, their livelihoods, and their social standing be destroyed by machines that could produce cloth faster, cheaper, and in larger quantities than any human weaver could match. Their protest was not against technology as such. It was against the specific distribution of costs and benefits that the technology produced: enormous profits for factory owners, destruction for skilled workers, and a transformation of the social order that concentrated wealth and power while distributing precarity and dependence.

The Luddites lost. The machines stayed. The factory system expanded. And the skilled craftsmen who once occupied respected positions in their communities were transformed, within a generation, into interchangeable components of a production system that valued their labor by the hour rather than by the skill.

This story resonates in the AI era for obvious reasons. The skilled knowledge workers who are currently watching AI systems approximate their capabilities — the lawyers, the writers, the designers, the analysts, the programmers, are in a structurally similar position to the hand-weavers of 1812. Their skills are real. Their expertise is genuine. Their concern about displacement is rational. And the standard response to their concern, "you're being a Luddite, technology always creates more jobs than it destroys" — may prove as inadequate as it was for the original Luddites, who did in fact lose their jobs, their skills, and their social position, regardless of whether the economy as a whole eventually created new forms of employment.

The lesson of the Luddites is not that resistance is futile. It is that the costs of technological change are real, are concentrated on specific populations, and are often dismissed by those who benefit from the change. The factory owner who called the Luddites backward was not wrong that the machines were more efficient. He was wrong that efficiency was the only thing that mattered.

There is a scene from the early industrial period that captures the psychological transformation with visceral clarity.

In 1830, the Liverpool and Manchester Railway began regular service — the world's first intercity passenger railway. On opening day, a Member of Parliament named William Huskisson was struck and killed by a locomotive during a stop at Parkside station. He had stepped onto the tracks to greet the Duke of Wellington, not realizing how quickly the engine was moving. He was accustomed to a world in which vehicles moved at the speed of horses. The locomotive moved at thirty miles per hour, a speed that his perceptual architecture, calibrated by a lifetime of horse-speed experience, could not accurately process.

Huskisson's death was not merely an accident. It was a collision between two worlds, the pre-industrial world of human-speed experience and the industrial world of machine-speed reality. His body had been trained by one set of physical assumptions. The machine operated by another. The gap between the two killed him.

This gap, between the pace at which human beings naturally process the world and the pace at which industrial systems operate, is one of the industrial revolution's defining features, and it has only widened with each subsequent technological acceleration. The human nervous system evolved for a world that moved at walking speed, seasonal pace, generational rhythm. The industrial world moved at engine speed, factory tempo, railway schedule. The digital world moves at processor speed, network pace, algorithmic tempo. At each step, the gap between human perceptual capacity and system speed has grown, and at each step, humans have adapted by developing new habits, new expectations, and new tolerances that bring them closer to the system's demands.

The factory taught workers to operate at machine speed. The railway taught travelers to perceive at vehicle speed. The telegraph taught communicators to expect messages at wire speed. Each adaptation was experienced not as a compression of the human but as an expansion of the possible. And each one left behind a residue: a slightly faster baseline expectation, a slightly lower tolerance for delay, a slightly narrower window of patience.

That cumulative narrowing of patience is the industrial revolution's most enduring psychological legacy. A person in 1800 who had to wait three weeks for a letter might have felt mild anticipation. A person in 1900 who had to wait three days for a telegram might have felt impatient. A person in 2000 who had to wait three hours for an email might have felt anxious. A person in 2024 who has to wait three seconds for a web page to load may feel frustrated enough to close the tab.

The objective duration of the wait has shortened by orders of magnitude. The subjective tolerance for waiting has shortened in parallel. This is not because modern people are constitutionally impatient. It is because the industrial revolution, and its digital successors, have systematically trained the nervous system to expect speed and to interpret delay as failure.

There is also the question of what the industrial revolution did to the human relationship with nature. Before industry, the rhythms of human life were, to a significant degree, synchronized with natural rhythms. Work happened during daylight. Rest happened during darkness. Activity varied with the seasons. The body's relationship with the environment was, if not harmonious, at least rhythmically connected.

Industry severed that connection. Artificial lighting extended the workday beyond sunset. Heated factories and workshops made seasonal temperature irrelevant to production schedules. The clock replaced the sun as the primary organizer of daily activity. Human beings, for the first time in evolutionary history, began living according to rhythms set by machines rather than by the natural world.

The consequences of this disconnection are still unfolding. Circadian rhythm disruption, seasonal affective disorder, vitamin D deficiency, the epidemic of sleep disorders in industrialized societies, all of these can be traced, at least in part, to the industrial revolution's severance of the ancient connection between human biology and natural rhythms. The modern world's growing interest in "reconnecting with nature", through outdoor recreation, meditation, circadian lighting, and digital detox, is, in this light, not a lifestyle trend. It is a symptom of a species that has been separated from the environmental signals its nervous system evolved to rely on.

Industry did not merely change how humans worked. It changed the environment in which human biology operates. And the mismatch between industrial environments and evolutionary expectations is one of the most underappreciated sources of modern psychological distress.

To see how radical this was, it helps to imagine a person moving from loosely coordinated labor into a factory system. Bells, shifts, repeated motions, overseen pace, standardized tasks, punctual arrival, punctual departure, quantified expectation. The body is no longer simply tired by labor. It is entrained by external tempo. That is a different kind of power.

There is a dimension of industrial time that connects directly to the digital present: the concept of productivity as a moral category.

Before industry, the word "productive" had primarily agricultural connotations, it described land that yielded crops, animals that bore young, seasons that brought harvests. The application of "productive" to human beings, the idea that a person could be more or less productive, that productivity was a measure of personal worth, that an unproductive hour was a wasted hour, is an industrial invention. And it is an invention so thoroughly absorbed that most modern people cannot imagine evaluating their own lives without it.

Picture the language modern professionals use to describe their days. "I had a productive morning." "I wasted the afternoon." "I need to be more efficient with my time." "I can't justify spending an hour on something that doesn't produce results." This language is not descriptive. It is evaluative, it applies a standard of judgment that equates human worth with measurable output. And the standard is so deeply internalized that it operates automatically, generating guilt when productivity falls and satisfaction when it rises, regardless of whether the activity in question has any intrinsic meaning or value.

This internalized productivity imperative is one of the industrial revolution's most significant psychological legacies. It persists long after the specific conditions that generated it, factory work, piece-rate payment, supervisor surveillance, have been transformed or eliminated. The modern knowledge worker who feels guilty for taking a lunch break, who checks email before bed, who measures their day in tasks completed rather than experiences had, is obeying a psychological program that was written by the factory system two centuries ago.

The digital economy has intensified this imperative by making productivity continuously measurable. In the industrial era, productivity was measured at the level of the shift, the day, or the week. In the digital era, it can be measured in real time — through keystroke logging, activity monitoring, task-completion dashboards, and the ambient pressure of asynchronous communication that creates a permanent record of responsiveness. The worker is never not measurable. And the measurement is never not happening.

This continuous measurement changes the experience of work in subtle but profound ways. The industrial worker knew when they were being observed and when they were not. The digital worker is always, potentially, being observed — through the traces their activity leaves in digital infrastructures. This is not necessarily sinister. Most digital monitoring is not conducted by malicious overseers. It is a structural consequence of working within systems that automatically record activity. But the effect on the worker's psychology is the same regardless of intent: the experience of being continuously accountable, continuously measurable, and continuously evaluated against a standard of productive output.

The result is a population that is, by historical standards, extraordinarily productive and extraordinarily stressed. The two are not contradictory. They are complementary, both products of the same psychological program, installed by the industrial revolution and amplified by the digital one: the program that equates human worth with measurable output and treats every moment of non-production as a moment of failure.

There is also the industrial revolution's effect on the concept of leisure. Before industry, leisure was not a distinct category. Work and rest intermingled throughout the day, shaped by the rhythms of the task and the body. The farmer worked hard during planting and harvest and rested during winter. The craftsman worked until the piece was done and then stopped. The boundary between work and not-work was fluid, permeable, and largely self-determined.

Industry hardened that boundary. Work happened during the shift. Leisure happened after. The weekend, a concept that did not exist in its modern form before the industrial era, was invented as a compensatory time-block: a period of officially sanctioned non-work designed to restore the worker's capacity for the next period of work. Leisure, in the industrial framework, is not an end in itself. It is a means, a mechanism for maintaining productive capacity.

This instrumental view of leisure persists in modern culture, often disguised as self-care. The person who meditates to improve focus, exercises to increase energy, sleeps to optimize performance, and vacations to prevent burnout is not, in the deepest sense, resting. They are maintaining the machine. They are performing what might be called "productive leisure" — activities whose value is measured not by the experience they provide but by the productivity they enable.

The concept of doing nothing — of rest without purpose, of time spent without any justification other than the pleasure of being alive — has become, in industrialized societies, almost impossible to articulate without apology. "I did nothing this weekend" is spoken with embarrassment, not satisfaction. The industrial revolution did not merely change how people work. It made not-working feel like a sin.

The factory is not only a place of production. It is a school for obedient timing.

This schooling extended far beyond factory walls. As industrial life spread, modern cities reorganized themselves around synchronized movement. Railways demanded precision. Schools trained punctuality. Offices absorbed schedule discipline. Entire populations became more legible to timekeeping. The industrial order made it easier for institutions to know not only where people were, but when they should be there.

The gains are obvious and immense. Industrialization increases output, lowers some costs, expands transportation, accelerates material production, and creates the conditions for many later improvements in wealth and living standards. But industry also intensifies alienation, exhausts bodies, fragments community, and normalizes the idea that life can be arranged around systems whose pace is not human by origin.

Here is one of the most haunting patterns in the book: architectures become powerful when human beings begin evaluating themselves by the standards those structures require.

Industry teaches this lesson with exceptional force. The disciplined worker becomes a social model. Regularity becomes moral. Timekeeping becomes identity.

The clock enters the soul when people stop experiencing industrial rhythm as external discipline and start experiencing it as the proper order of life.

Here is normalization at full power.

It also changes the meaning of labor. Work becomes increasingly abstracted into hours, shifts, and outputs. One does not merely make or tend or carry out a task in concrete relation to its context. One performs labor as measurable production. Human beings begin seeing their own time as divisible units that can be bought, sold, tracked, and compared.

That shift is not trivial. It makes wage labor more scalable, organizations more manageable, and modern capitalism more psychologically sustainable. It also deepens estrangement between activity and meaning. If a person's worth is increasingly evaluated by productivity under externally defined conditions, then the person becomes easier to organize but harder to feel whole.

Industry therefore does something more than accelerate material life. It produces a new standard human.

This human is punctual, productive, internally monitored, externally timed, and accustomed to living inside large structures that value consistency over spontaneity. Such a person may gain wages, urban access, mobility, and social participation in modern economies. But such a person also absorbs a deeper lesson: the body exists under schedules it did not author.

This pattern will recur later under digital systems, where the demand is no longer only physical punctuality but cognitive and attentional responsiveness. In that sense, industry is one of the great ancestor chapters of contemporary life. It normalizes externally organized human rhythm.

The machine does not need to dominate by spectacle. It dominates by cadence.

By the time industrial discipline matures, it no longer looks like an unusual intervention into human life. It looks like adulthood, employment, seriousness, and normal order. The historical shock disappears. The clock remains.

And once time is disciplined, another frontier opens.

There is a detail from the early factory era that captures the violence of the temporal transformation with startling precision. In many early textile mills, the factory clock was the only clock in the community. Workers who arrived at the factory gate had no independent means of verifying the time. Factory owners were known to manipulate the clock, advancing it in the morning to extend the work period and retarding it in the evening to delay the end of the shift. The workers, who had no watches of their own, had no recourse.

This manipulation was possible because the factory had achieved a monopoly on time itself. The owner did not merely control the means of production. He controlled the means of temporal measurement. The workers lived inside his time, on his schedule, according to his clock. Their experience of duration, of when the day began and ended, of how long they had worked, was determined by an instrument they did not own and could not verify.

The parallel to the digital age is direct. The platforms that structure modern work and social life control the metrics by which experience is measured. Screen time reports, engagement metrics, productivity dashboards, algorithmic feeds — all of these are temporal instruments controlled by entities whose interests may not align with the user's. The user who checks their screen time report is consulting a measurement produced by the same entity that designed the environment the measurement describes. They are, like the factory worker checking the factory clock, relying on the system itself to tell them how much of their life the system has consumed.

Attention itself can be organized.

Benjamin Franklin popularized the phrase "time is money" in 1748 — a period when the transition from artisanal to industrial rhythms was underway. The phrase captures an entire worldview: that time has no intrinsic value except as productive output, that unused time is wasted, and that the proper relationship between person and hours is one of optimization.

This worldview is now so embedded that questioning it feels almost perverse. What else would time be for? The answer — that time might be for rest, contemplation, play, or the simple experience of being alive without producing, sounds like luxury or moral failure. That reaction is itself evidence of how completely the industrial relationship with time has been absorbed.

The costs are visible in modern epidemiology. Chronic stress, burnout, sleep deprivation, and the vague sense of never having enough time are among the most common complaints in affluent societies — societies where, by historical standards, people have more comfort, more technology, and more leisure than any previous generation. The paradox makes sense only through the industrial revolution's legacy: the clock reorganized the experience of time itself, making every hour accountable, every idle moment guilty.

Industry also transformed childhood. Before industrialization, children were integrated into household economic life. The industrial revolution first exploited this (child labor in factories was among the era's worst atrocities) then replaced it with compulsory schooling. The modern school — with timetables, bells, age-based cohorts, standardized curricula, was designed to produce an industrial workforce. The school day mirrors the factory day. The school bell mirrors the factory bell. The emphasis on following instructions, sitting still, and completing tasks on schedule mirrors industrial labor requirements. This is documented educational history: the Prussian model of compulsory education was explicitly designed to produce citizens literate enough to follow instructions and disciplined enough to submit to authority.

And once rhythm has been industrialized, the next question becomes: can influence be industrialized as well? If the factory disciplined the body, could a new technology discipline attention itself?

The answer arrives through signal.

Electricity and mass media follow.

CHAPTER 10

Electricity and Mass Media: When Influence Became Ambient

Electricity changed modern life by extending activity beyond older limits of daylight and locality.

Picture a family gathered around a radio, hearing the same voice at the same moment as millions of others they will never meet. That scene contains one of the central secrets of modern influence: intimacy can now be manufactured at scale. But electricity alone is only half the story. The deeper civilizational shift arrives when electrified media begin shaping perception at scale.

At that point, influence becomes ambient.

A message no longer travels only by person, paper, or local gathering. It can enter homes simultaneously. It can synchronize attention across vast distances. It can bind populations into common emotional rhythms.

The result is a profound transformation. The public becomes more than a physical crowd. It becomes an electronically coordinated field of shared images, voices, fears, and desires.

Radio, film, and television each intensify this in different ways. They do not merely inform. They stage reality. They do not merely entertain. They normalize common atmospheres of feeling.

A nation can now hear itself.

A market can now seduce itself.

A regime can now persuade at scale.

Mass media therefore belong to this book not only as communication tools, but as emotional infrastructures. They teach human beings to inhabit mediated worlds so regularly that mediation itself stops feeling unusual.

One no longer needs direct proximity to participate in a collective mood. One can be moved by distant voices and distant images repeatedly enough that the mediated world begins to compete with local reality for authority.

The gains are significant. Shared information expands. Cultural reach grows. Entertainment becomes widely accessible. Political messages can mobilize populations. Education can travel. But the costs follow the familiar pattern. Influence grows more atmospheric, more centralized, more repeatable, more intimate.

Persuasion no longer always appears as argument. It appears as environment.

This matters because environments shape people most effectively when they are no longer experienced as interventions.

By the time mass media maturity is reached, whole societies are accustomed to being guided by recurring symbols, narratives, advertising cues, and emotionally charged simplifications. Modern propaganda and modern consumer culture are cousins because both understand the same principle: repetition can become reality's escort.

The public mind becomes easier to coordinate when the instruments of attention are centralized.

Electricity matters here because it alters continuity itself. Darkness no longer sets the same hard boundary. Activity extends. Consumption extends. Signal extends. The day becomes more permeable to intervention. This does not create mass media alone, but it creates a world in which mediated rhythms can remain active with unprecedented persistence.

The modern mind therefore does not just learn to watch. It learns to live inside atmospheres generated elsewhere.

The most unsettling feature of this transition is that the influence often feels voluntary. A person sits down to be informed or entertained. Yet the medium is simultaneously teaching tempo, framing emotion, repeating symbols, narrowing attention, and establishing what belongs in public imagination. Over time, mediated presence no longer feels like an interruption of ordinary life. It feels like part of ordinary life itself.

Here is the normalization.

Electricity and mass media also alter simultaneity. Millions can now occupy the same emotional moment without sharing physical space. This matters politically as much as culturally. Fear can be synchronized. Hope can be synchronized. Consumption can be synchronized. Outrage can be synchronized. A society begins experiencing itself through recurring electronic mirrors.

Those mirrors do not simply reflect public feeling. They help produce it.

This is why the chapter belongs between industry and computing. Industry standardizes bodies around time. Mass media standardize populations around mood and attention. The body is already being trained; now the atmosphere is trained around it.

This creates a new kind of passivity and a new kind of power. When influence becomes ambient, it does not always need to persuade through explicit argument. It can shape through repetition, familiarity, prestige, and emotional association. The citizen becomes audience. The audience becomes market. The market becomes a daily pedagogy of desire.

There is a specific moment in the history of mass media that captures the transformation with unusual precision: the introduction of the laugh track.

In 1950, the sound engineer Charley Douglass invented a device he called the "laff box", a keyboard-controlled machine that could produce the sound of audience laughter on cue, allowing television producers to add laughter to comedies that were filmed without a live audience. The device was crude by modern standards, but its effect was transformative. Shows that used the laugh track were perceived as funnier by audiences. The presence of recorded laughter — canned, artificial, clearly not generated by anyone in the viewer's living room, somehow made the jokes land better.

This is a small, seemingly trivial innovation. But it reveals something profound about the relationship between media and psychological experience. The laugh track works because human beings are social creatures whose emotional responses are influenced by the perceived responses of others. If you hear other people laughing, you are more likely to laugh. If you hear silence, you are more likely to judge the joke as unfunny. The laugh track exploits this social-emotional mechanism by simulating the presence of an amused audience, an audience that does not exist, that is not watching the show, that is, in fact, nothing more than a recording of anonymous laughter produced by a machine.

And yet it works. It works despite the fact that every viewer knows, at some level, that the laughter is fake. The knowledge does not prevent the effect. The emotional circuitry responds to the behavioral cue, the sound of laughter — regardless of the rational mind's assessment of its authenticity.

This is precisely the mechanism that will later be exploited by social media (where the "like" count functions as a social proof signal analogous to the laugh track), by recommendation systems (where "people who bought this also bought..." functions as simulated social endorsement), and by AI-generated content (where fluent, confident prose functions as a signal of expertise regardless of whether any expertise was involved in its production). In each case, the arrangement produces a behavioral cue that the human emotional system interprets as social evidence, and the interpretation occurs automatically, below the level of conscious evaluation, regardless of whether the person "knows" the cue is manufactured.

The laugh track is, in miniature, the entire pattern of media manipulation that this chapter describes: the construction of emotional environments through artificial signals that exploit the automatic, socially calibrated responses of the human nervous system. The sophistication of the signals has increased enormously since 1950. The underlying mechanism has not changed at all.

There is a darker consequence here as well. When public feeling is mediated often enough, people begin confusing visibility with reality. What is seen repeatedly feels socially larger than what is merely lived quietly. The exceptional event outcompetes the ordinary condition. Drama acquires advantage over proportion. This does not begin with social media. It begins earlier, in the age when image and voice first learn to arrive together at scale.

That shift changes politics, but it also changes the self. Individuals begin measuring their place in the world against mediated archetypes: the successful face, the desirable body, the patriotic image, the feared outsider, the admired lifestyle, the glamorous purchase, the acceptable emotion. Mass media do not invent aspiration or anxiety. They amplify both by making comparison ambient.

This is why electrified media belong in the psychological history of normalization. They teach societies not only what to notice, but what kinds of feeling are publicly available to them. They can make grief national, panic contagious, glamour aspirational, war intimate, and consumption emotionally charged.

Once that machinery matures, the person is no longer merely receiving information. The person is living inside a curated weather system of symbols.

And once attention can be shaped at scale, the next transition is not merely toward more content, but toward more processing.

There is a specific technique from the history of advertising that reveals the structure of media influence with particular clarity: the concept of "manufactured dissatisfaction."

In the 1920s, the advertising industry underwent a transformation from informational advertising (which described the features of products) to psychological advertising (which created emotional needs that products could satisfy). The pioneer of this approach was Edward Bernays, Sigmund Freud's nephew, who applied his uncle's insights about unconscious desire to the problem of selling consumer goods.

Bernays understood that the most effective advertising does not merely describe a product. It creates a problem that the product solves. The problem need not be real. It need only be felt. An advertisement that makes a woman feel self-conscious about her appearance has created a problem. The product it sells is the solution. An advertisement that makes a man feel inadequate about his car has created a problem. The new model is the solution. The genius of psychological advertising is that it manufactures the very dissatisfaction it promises to relieve.

This technique, refined over a century of practice, is now embedded in the architecture of every digital platform. Social media does not merely display content. It creates comparison. The feed does not merely show what others are doing. It makes you feel that what you are doing is insufficient. The recommendation does not merely suggest a product. It implies that your current life lacks something. The notification does not merely inform. It creates the anxiety that something important is happening without you.

The advertising industry invented manufactured dissatisfaction as a commercial technique. Digital platforms have automated it as an architectural principle. The structure is identical. Only the scale has changed.

The machine will begin handling logic itself.

There is a phenomenon called "mean world syndrome", a term coined by researcher George Gerbner, describing heavy television viewers' tendency to perceive the world as more dangerous than it actually is. Gerbner found that people who watched more TV were more likely to overestimate violent crime risks and believe most people cannot be trusted. The mechanism is straightforward: TV news disproportionately covers dramatic, threatening events. A viewer absorbing hours of coverage develops a model of reality that is systematically more frightening than direct experience would produce.

There is one more phenomenon associated with mass media that this chapter must address: the creation of what the cultural theorist Guy Debord called "the society of the spectacle."

Debord's argument, published in 1967, was that modern societies had undergone a fundamental transformation in the relationship between experience and representation. In pre-media societies, people's primary experience of the world was direct: they saw with their own eyes, heard with their own ears, and formed judgments based on personal encounters. Mass media reversed this relationship. Increasingly, people's primary experience of the world was mediated, filtered through images, narratives, and representations produced by institutions whose interests did not necessarily align with those of the audience.

The "spectacle," in Debord's formulation, is not merely entertainment. It is a mode of social organization in which representations substitute for reality, in which images of life replace the experience of life, and the consumption of representations becomes the primary form of social participation. The person who watches a cooking show rather than cooking, who follows travel influencers rather than traveling, who engages in political debate on social media rather than participating in community governance, is not merely being entertained. They are inhabiting a world in which mediated experience has become primary and direct experience secondary.

Debord's analysis was prescient but also incomplete. He wrote in an era of broadcast media, when the spectacle was produced by a relatively small number of institutions and consumed by a relatively passive audience. The digital age has democratized the production of spectacle, anyone with a smartphone can create and distribute images, narratives, and representations to a global audience. But democratization has not resolved the underlying problem. It has intensified it. The volume of mediated experience has increased exponentially, while the capacity for direct, unmediated experience has, if anything, decreased.

Consider what it means to live in a world where the average person spends several hours per day consuming media of various kinds — scrolling social feeds, watching streaming content, reading news, listening to podcasts, participating in online discussions. Each hour spent consuming media is an hour not spent in direct sensory engagement with the physical world, in face-to-face conversation with other human beings, or in the kind of unstructured mental activity (daydreaming, reflection, imagination) that psychologists associate with creativity and self-understanding.

Here is not a moral judgment. Much of what media provides is genuinely valuable — information, education, connection, entertainment, exposure to perspectives one would never encounter locally. The question is not whether media is good or bad. The question is what happens to a species whose primary mode of experiencing the world shifts from direct engagement to mediated consumption. And the answer, as with every revolution in this book, is that the shift changes what feels normal — until the mediated world feels more real, more vivid, more engaging than the unmediated one.

That inversion, when the representation becomes more compelling than the reality it represents, is one of the defining psychological conditions of late modernity. It was established by broadcast media. It was intensified by the internet. It will be further intensified by

AI-generated content that is indistinguishable from human production and potentially more engaging than anything a human could create.

The society of the spectacle did not end with Debord. It evolved. And its next evolution may be the most far-reaching yet: a spectacle generated not by human institutions but by machines, optimized not for human understanding but for human engagement, and indistinguishable from reality not because it tries to deceive but because it has become better than reality at holding attention.

This matters because it illustrates a general principle: media environments do not merely transmit information. They construct the world as experienced by their audiences. And the media-constructed world does not feel constructed. It feels like the world. The heavy viewer does not think "my perception of danger may be inflated." They think "the world is dangerous." The medium's influence is invisible because it is pervasive.

This dynamic intensified with each subsequent revolution. The internet amplified it by removing gatekeepers. Social media amplified it with algorithmic optimization. AI amplified it by making persuasive content nearly costless. But the template was established by broadcast media.

There is a phenomenon that became visible in the age of television and has intensified in every subsequent medium: the displacement of firsthand experience by mediated experience as the primary basis for belief, opinion, and emotional response.

Picture how a person in a pre-media world forms an opinion about a distant event. They hear about it from a traveler, from a letter, from a town crier. The information is sparse, delayed, and filtered through human intermediaries whose biases and limitations are at least partly visible. The person knows they are receiving a secondhand account. They know that the account may be incomplete, colored by the teller's perspective, or simply wrong. They maintain, almost automatically, a degree of epistemic humility, a recognition that their knowledge of distant events is limited and provisional.

Now consider how a person in a television-saturated world forms an opinion about a distant event. They see footage. They hear eyewitness accounts. They watch experts analyze the situation in real time. The information is vivid, immediate, and emotionally compelling. The person feels, and this is the critical word, that they have witnessed the event. They have seen it. They know what happened. The epistemic humility that characterized secondhand knowledge evaporates, replaced by the conviction that direct visual exposure constitutes understanding.

But visual exposure is not understanding. A thirty-second news clip of a complex geopolitical event does not provide understanding. It provides an emotional impression, a feeling of knowing that substitutes for the slow, difficult, contextual work of actually understanding. The viewer who has watched the clip feels informed. They may, in fact, be misinformed, not by deliberate falsehood, but by the structural limitations of a medium that presents complex events as simple visual narratives.

This displacement of understanding by the feeling of understanding is one of mass media's most important effects. It produces a population that is, by every metric, more exposed to information about the world than any previous generation — and that may, for that very reason, be less capable of the kind of slow, contextual, humble reasoning that understanding requires. The media-saturated mind is not empty. It is full — full of impressions, images, emotional reactions, and half-formed opinions that feel like knowledge because they arrived through a channel that mimics the directness of personal experience.

Television created this dynamic. The internet intensified it by adding interactivity, personalization, and the illusion of participation. Social media intensified it further by making every user simultaneously a consumer and producer of pseudo-knowledge. AI will intensify it again by enabling the generation of persuasive, emotionally compelling content at a scale and speed that makes evaluation essentially impossible.

The trajectory is clear: each successive medium increases the volume and vividness of mediated experience while decreasing the time and cognitive resources available for evaluation. The result is a civilization that knows more and understands less, that is drowning in information and starved of wisdom.

The result is not merely a metaphor. It is a description of a specific cognitive condition: the state of being overwhelmed by inputs that the mind cannot process at the rate they arrive. Psychologists call it "information overload." Neuroscientists call it "cognitive overload." Ordinary people call it "feeling overwhelmed." Whatever the label, the experience is the same: a mind that has more to process than it can handle, and that copes not by processing more deeply but by processing more superficially — skimming, scrolling, reacting rather than reflecting.

Mass media created the conditions for this overload. Every subsequent medium has deepened it. And the deepening continues.

There is one final dimension of mass media's revolution that connects directly to the digital age: the creation of celebrity culture and its effects on ordinary self-perception.

Before mass media, fame was local. A person might be known throughout their village, their town, perhaps their region. But the scale of renown was limited by the reach of physical communication. Mass media changed this by creating a new category of social existence: the celebrity — a person known by millions who know nothing of them in return.

Celebrity is a one-directional social relationship. The celebrity is visible; the audience is invisible. The celebrity performs; the audience watches. The celebrity's life is documented, discussed, analyzed, and consumed by millions of people who will never meet them, never speak to them, never enter their world. This asymmetry is historically unprecedented. For most of human existence, social knowledge was reciprocal: if you knew someone's name, they knew yours.

The psychological effects of living in a world populated by celebrities are more significant than they may first appear. The human brain did not evolve to distinguish between people known through media and people known through direct interaction. Research in social comparison theory, beginning with Leon Festinger's work in the 1950s and extending through decades of subsequent study, shows that people evaluate their own lives, their appearance, their success, their relationships, their worth, partly by comparison with the people they see around them.

In a world without media, the comparison set is local: your neighbors, your colleagues, your community. The people you compare yourself to are, in most respects, not dramatically different from you. In a world of mass media, the comparison set expands to include celebrities, models, athletes, entrepreneurs, influencers, and curated representations of extraordinary lives. The person you compare yourself to is no longer your neighbor. It is a carefully managed image of someone whose life has been edited, filtered, lit, and staged for maximum impact.

The result is a systematic distortion of self-perception. Studies have consistently found that exposure to idealized media images is associated with lower body satisfaction, lower self-esteem, and higher rates of anxiety and depression, particularly among young people and women. This is not because people are foolish enough to believe that media images represent reality. It is because the comparison process operates automatically, below the level of conscious evaluation. The brain compares, registers a deficit, and generates a negative emotional response — all before the rational mind has had time to remind itself that the comparison is meaningless.

Social media has intensified this dynamic by democratizing the production of curated self-presentation. On platforms like Instagram,

TikTok, and their successors, every user is simultaneously a consumer and a producer of edited images of life. The average person now curates their own appearance, their own experiences, and their own life narrative for public consumption, and consumes the curated presentations of hundreds or thousands of others. The comparison process, which mass media initiated with celebrities, now operates continuously among ordinary people, each one simultaneously the audience and the performer in an unending theater of social evaluation.

The psychological cost is measurable. Studies of social media use consistently find associations between heavy use and negative mental health outcomes, particularly among adolescents. The mechanisms are consistent with social comparison theory: exposure to curated images of others' lives generates a sense of inadequacy that is resistant to rational correction because it operates at the level of automatic emotional response rather than conscious evaluation.

Mass media created the template for this dynamic by introducing the celebrity — the universally visible, unilaterally known social figure against whom ordinary people measured themselves. Social media democratized the template. AI will soon be capable of generating unlimited quantities of photorealistic images of people who do not exist, in situations that never occurred, with bodies that were never born, images that are optimized, through algorithmic learning, to produce maximum engagement, maximum comparison, and maximum dissatisfaction.

The trajectory is clear: from celebrity culture to social media curation to AI-generated social comparison. At each step, the distance between the image and reality grows, the volume of images increases, and the psychological impact of comparison intensifies.

Mass media did not merely change what people saw. It changed how people saw themselves. And the change persists, deepens, and accelerates with every subsequent medium.

Mass media also introduced the commodification of attention. Before mass media, attention was a private resource directed by individual interest. Mass media transformed attention into an economic commodity. The broadcast model — free content funded by selling audience attention to advertisers — creates structural incentive to maximize captured attention. This incentive shaped every subsequent medium: newspaper front pages, radio formats, TV scheduling, website layouts, social media feeds, recommendation algorithms. In each case, the medium's economic logic works against the user's cognitive autonomy. The medium needs your attention to survive. You need your attention to think.

PART IV

The Collapse of Inner Distance

The relationship between mass media and democracy deserves particular attention, because it reveals one of the deepest tensions in the modern world: the tension between informed citizenship and manipulated attention.

Democratic theory depends on an informed citizenry, people who have access to accurate information about public affairs and the cognitive capacity to evaluate that information critically. Mass media, in principle, serves this function: it provides citizens with information about events, policies, and leaders that they could not observe directly. In practice, the relationship is far more complicated.

The problem is not that mass media lies (though it sometimes does). The problem is structural. A medium funded by advertising must compete for attention. Competition for attention rewards the dramatic, the emotional, the threatening, the novel, and the polarizing. Nuance, complexity, historical context, and careful qualification — the qualities most needed for informed democratic deliberation, are precisely the qualities that perform worst in an attention economy. A headline that says "New Study Shows Modest Improvement in One Metric Under Certain Conditions" will never compete with "Crisis Deepens as Expert Warns of Catastrophe."

This structural bias does not require bad faith from journalists. Many journalists are deeply committed to accuracy and nuance. But they work within institutions that are subject to economic pressures that systematically reward simplification and sensationalism. The result is a media environment that is simultaneously more informative and more distorting than any previous communication structure — providing unprecedented access to information while systematically shaping how that information is framed, prioritized, and emotionally colored.

The political consequences are visible in every modern democracy. Voters who are, in principle, better informed than any previous generation in human history are simultaneously more susceptible to manipulation, more polarized in their views, and more likely to hold demonstrably false beliefs about basic facts. This paradox makes sense only when understood through the lens of the media environment: people are not uninformed. They are misinformed, shaped by a communication system whose economic incentives are misaligned with the cognitive requirements of democratic citizenship.

This problem did not begin with the internet. It began with broadcast media. And understanding its origins in radio and television is essential for understanding its intensification in the digital age.

The transition from broadcast to digital media did not introduce a new problem. It accelerated and personalized an old one. Broadcast media distorted public attention through editorial choices: what to cover, how to frame it, how much time to give it. Digital media distorts public attention through algorithmic choices: what to surface, what to suppress, what to recommend, and — crucially — what to optimize for. The shift from editorial curation to algorithmic optimization is a shift from human judgment (flawed but accountable) to machine logic (efficient but opaque). The result is a media environment that is even more effective at capturing attention and even less aligned with the requirements of informed citizenship.

Marshall McLuhan's famous dictum — "the medium is the message" — is often quoted and rarely fully understood. McLuhan's point was not that content does not matter. His point was that the *form* of a medium shapes cognition and culture more powerfully than any particular content transmitted through it. Television, regardless of what it shows, trains the eye to expect visual stimulation and the mind to expect emotional engagement. The internet, regardless of what it contains, trains attention to expect speed, novelty, and interactivity. The smartphone, regardless of what apps are installed, trains the nervous system to expect constant availability and instant gratification.

These are not effects of bad content. They are effects of the medium itself. And they cannot be corrected by producing better content, any more than the health effects of a sedentary lifestyle can be corrected by sitting in a more ergonomic chair. The problem is not what the medium carries. The problem is what the medium does to the minds that inhabit it.

Modern digital systems do more than connect. They enter the structure of daily life, compressing the distance between signal and response, public and private, system and self. In this part, abstraction becomes intimate.

The key shift here is not only technical. It is psychological. The system is no longer merely outside the person, waiting to be used. It begins living inside routine, habit, attention, and expectation.

CHAPTER 11

Computers: When Logic Left the Human Hand

The computer often appears in popular imagination as an object of innovation, speed, and convenience.

Long before it felt conversational or creative, however, it had already transformed civilization in a quieter way: by teaching institutions to trust machine-handled logic in places where scale had outgrown ordinary human processing.

Before widespread computing, many forms of complexity remained limited by human processing capacity. Records could be kept, but not at the same scale. Patterns could be examined, but not at the same speed. Decisions could be organized, but not with the same computational assistance. The computer changes this by making symbolic operations executable outside the human mind and hand.

What emerges is a subtle but civilizationally enormous shift.

A machine can now sort, count, match, model, store, retrieve, and calculate at speeds and scales ordinary human cognition cannot sustain unaided. In the beginning this often looks administrative rather than existential. Payroll, census, logistics, accounting, inventory, bureaucracy. Yet that is precisely where its power hides.

Human beings first normalized machine logic through paperwork.

Imagine the quiet authority of the form. A box to tick. A field to complete. A number to enter. A record to retrieve. Civilization does not always feel itself changing through spectacle. Sometimes it changes through the spread of process. The filing cabinet, the database, the structured record, the coded entry --- these do not merely organize information. They teach the world to become easier for systems to read.

The filing cabinet becomes a threshold to a new world. Data frameworks quietly replace memory. Procedure migrates into software. Institutions begin trusting outputs they cannot fully inspect line by line. The system becomes a partner in interpretation.

Here is not yet artificial intelligence in the contemporary sense, but it is already a major concession: the world will increasingly be managed through layers of machine-mediated logic.

The gains are enormous. Complex organizations become more possible. Scientific modeling improves. Administration scales. Information storage expands. Planning accelerates. Yet a familiar dependency enters. The more life is organized through systems too complex for ordinary intuition, the more trust shifts away from direct understanding and toward technical mediation.

That mediation slowly becomes normal.

At first, the computer is a specialized machine. Then it becomes a work necessity. Then a cultural necessity. Then an everyday background. What once seemed remarkable becomes infrastructure.

There is one more observation about the computer revolution that connects it directly to the AI revolution described in later chapters: the computer taught civilization to trust processes it could not understand.

The point is not an obvious point, because we are accustomed to thinking of computers as tools that we use. But the relationship between human users and computer systems has, from the beginning, been characterized by a fundamental asymmetry of understanding. Most people who use computers, who check email, browse the web, process spreadsheets, file taxes online, make payments, and manage their social lives through digital platforms — do not understand how computers work at even the most basic level. They do not know what an algorithm is. They do not know how data is stored, transmitted, or processed. They do not know what happens between the moment they click a button and the moment the screen changes.

And this does not bother them. They trust the network to work. They trust the output to be correct. They trust the process to be reliable. And their trust is, for the most part, well-founded, computers are extraordinarily reliable, and the systems built on them function smoothly enough that most people interact with them for years without encountering a failure serious enough to shake their confidence.

But this trust has a psychological consequence that is easy to miss: it habituates people to the experience of depending on systems they do not understand. Before computers, most of the systems that governed daily life were at least somewhat transparent. A farmer understood the basic principles of crop growth. A shopkeeper understood the basic principles of commerce. A citizen understood the basic principles of law. The systems might be complex, but they were built from components that were, in principle, comprehensible to an educated person.

The computer introduced opacity as a permanent feature of civilizational infrastructure. The average person has no more understanding of how their smartphone processes a search query than a medieval peasant had of how the sun moved across the sky. The difference is that the medieval peasant at least had a narrative explanation (the sun was carried by a celestial mechanism, or by the will of God). The modern user has no explanation at all, merely a vague confidence that "it works" and a willingness to trust the output without understanding the process.

This willingness to trust opaque frameworks is the precondition for everything described in the remaining chapters of this book. Algorithmic recommendation works because people trust the output without understanding the algorithm. Predictive policing works because institutions trust the data without understanding the model. AI-generated content works because users trust the fluency without understanding the mechanism. In each case, the trust was established not by the AI revolution but by the computer revolution — by decades of interaction with systems that worked reliably without being understood.

The computer taught civilization a habit that AI will exploit: the habit of treating outputs as trustworthy without demanding transparency about the process that produced them. That habit, more than any specific technology, may prove to be the computer revolution's most important legacy.

This chapter matters because it shows how human beings begin trusting non-human procedures without requiring those procedures to feel alive. The path to later machine authority does not begin with humanoid robots or sentient software. It begins with systems that are useful enough to be obeyed.

There is a dimension of the computer revolution that becomes visible only in hindsight: the way computers changed the concept of truth.

Before computers, truth was primarily a property of statements — sentences that could be evaluated as accurate or inaccurate descriptions of reality. "The bridge is 500 meters long" was true or false depending on the actual length of the bridge. "The patient has pneumonia" was true or false depending on the patient's actual condition. Truth was a relationship between language and the world it described.

Computers introduced a new concept of truth: computational validity. A program that runs without errors, that processes inputs according to its instructions and produces outputs consistent with its logic, is "correct" in a computational sense — regardless of whether its outputs correspond to external reality. A model that processes data and generates a prediction is "valid" if the mathematics are correct, regardless of whether the prediction turns out to be accurate. The computer's standard of truth is internal consistency, not external correspondence.

This distinction matters because it shapes how institutions evaluate claims. In a pre-computational world, a claim's credibility depended on its relationship to observable reality: can it be verified? Does it match what we see? In a computational world, a claim's credibility increasingly depends on its relationship to a model: was the data processed correctly? Are the algorithms sound? Is the output consistent with the inputs?

These are not the same question. A model can be computationally perfect and empirically wrong. A prediction can be mathematically valid and factually incorrect. The computer's standard of truth, internal consistency — can diverge from reality's standard of truth, external correspondence, and the divergence may not be visible to anyone who lacks the technical expertise to evaluate the model.

What emerges is one of the most decisive epistemological shifts of the computer age. It means that an increasing number of institutional decisions, about credit, employment, health, education, criminal justice, are based not on direct observation of reality but on the outputs of computational models whose relationship to reality is mediated by assumptions, data choices, and algorithmic logic that most people cannot evaluate.

The person denied a loan is told that their credit score is too low. But they cannot evaluate the model that produced the score. The patient recommended a treatment plan is told that the algorithm suggests this course. But they cannot evaluate the data on which the algorithm was trained. The citizen affected by a policy decision is told that the analysis supports this approach. But they cannot evaluate the model that generated the analysis.

In each case, truth has been delegated to a computational process, and the person affected by that truth has no independent means of verifying it. They must trust the output. And the output is presented not as the product of a model with specific assumptions and limitations, but as a fact — a number, a recommendation, a score — that carries the authority of mathematical precision regardless of its actual relationship to reality.

The first surrender to machine cognition is not emotional. It is administrative.

This matters more than it first appears. Administration is where civilizations hide some of their deepest assumptions about reality. What counts, who counts, what is tracked, what is ignored, what is categorized, what is processed, what is retrievable, what is official --- these are not trivial questions. Once computers begin governing those operations, they reshape how institutions think.

The human operator is still present, but the operator increasingly works through machine-compatible formats. Categories must fit. Inputs must be defined. Records must be standardized. The computer does not merely accelerate thought. It encourages the world to become more processable.

That encouragement is one of the most underappreciated forms of technological power.

A world that is easier for systems to process gradually becomes a world arranged for systems.

People adapt to forms, fields, databases, workflows, permissions, and digital procedures. Such adaptation often feels ordinary because it is practical. But practical adaptation is still adaptation. Human behavior bends toward machinic legibility.

This is the bridge between administrative computing and later forms of algorithmic life. Before the machine predicts, it organizes. Before it recommends, it structures. Before it appears intelligent, it teaches institutions to depend on machine-handled logic.

That dependency becomes cultural long before it becomes dramatic.

There is a mood to this shift that deserves attention. The computer age does not initially announce itself as a challenge to the human. It arrives as relief from clerical burden. It promises fewer errors, faster retrieval, cleaner organization, more scalable coordination. It helps the office, the agency, the enterprise, the laboratory, the state. It appears as competence in machine form.

That is why it is so easy to welcome.

But competence, once externalized, changes expectation. The worker accustomed to paper delay begins expecting instant retrieval. The institution accustomed to human approximation begins preferring structured precision. The manager accustomed to partial visibility begins expecting dashboards. The system does not merely save time. It alters what slowness, ambiguity, and missing information come to mean.

This is one of the deeper psychological preparations for the digital age. People become less tolerant of opacity once systems can reduce it. They become less tolerant of inconsistency once systems can standardize around it. In that sense, computing does not simply increase efficiency. It changes the emotional threshold of acceptable disorder.

That threshold shift matters because human life contains forms of ambiguity that do not yield gracefully to machine structure. Yet the more civilization becomes organized through computable formats, the more non-computable reality starts looking like noise, error, inefficiency, or risk.

The result is not the same as saying computers are dehumanizing by nature. It is saying they reward a certain style of legibility. Under their influence, institutions become more likely to value what can be entered, tracked, compared, and processed. The unstructured human remainder does not disappear, but it becomes harder to administer without pressure to convert it into data.

There is one more observation about the computer revolution that connects it to this book's deepest theme: the computer's role in creating the concept of the "user."

Before computers, people interacted with tools. A carpenter used a saw. A farmer used a plow. A writer used a pen. In each case, the relationship between the person and the tool was one of active manipulation: the person controlled the tool, directed it, and determined its operation through physical skill and conscious intention.

The computer introduced a different relationship: the relationship of the "user." A user does not control the computer in the way a carpenter controls a saw. A user operates within an environment that the computer provides — an environment designed by someone else, governed by rules the user did not write, and offering capabilities that the user can access but not fundamentally alter. The user's freedom is real but bounded: free to choose among the options the network provides, but not free to create options the system does not support.

This distinction — between controlling a tool and operating within a mechanism, is one of the most important psychological shifts of the computer age. A tool extends the user's will. A framework constrains the user's will within a designed environment. Both are useful. But the psychological experiences they produce are fundamentally different. The person who controls a tool experiences agency. The person who operates within a system experiences convenience.

The language is revealing. We do not "control" our computers. We "use" them. We do not "direct" our phones. We "navigate" them. We do not "command" our apps. We "interact" with them. The shift from active verbs of control to passive verbs of interaction reflects a genuine change in the relationship between human beings and their technologies, a shift from dominance to collaboration, and sometimes from collaboration to dependence.

This shift accelerated with each successive generation of computing technology. The mainframe user operated within a highly constrained environment, following rigid procedures to accomplish specific tasks. The personal computer user had more freedom but still operated within the constraints of an operating system and software applications designed by others. The smartphone user operates within an even more designed environment — one optimized not merely for functionality but for engagement, retention, and data extraction. The AI user may eventually operate within an environment so sophisticated that the boundary between the user's intentions and the system's suggestions becomes difficult to locate.

At each stage, the environment becomes more capable and more constraining simultaneously. More capable because the tools available to the user are more powerful. More constraining because the environment within which those tools operate is more thoroughly designed, designed not by the user but by entities whose interests may not align with the user's own.

The concept of the "user" is, in this sense, one of the computer revolution's most consequential innovations. It created a new social role, neither master nor servant, neither controller nor controlled, but something in between: a person who operates within a designed environment and experiences that environment as both empowering and pre-structured. The user is free — free within limits that they did not set and often cannot see.

That is one of the most precise descriptions of the human condition in the digital age. And it is a condition that the computer revolution created, that the internet expanded, that the smartphone made permanent, and that AI may eventually make inescapable.

There is one more dimension of the computer revolution that deserves attention: its effect on the concept of error.

Before computers, error was human. A clerk who miscalculated a sum, a typist who misspelled a word, a manager who misread a report, these were understood as failures of individual competence or attention. Error was personal, attributable, and correctable through training, discipline, or replacement of the erring individual.

Computers introduced a new category of error: mechanism error. When a computer crashes, when a database corrupts, when an algorithm produces an unexpected output, the error is not attributable to any individual human being. It is a property of the system, a failure in the interaction between code, hardware, data, and the countless environmental factors that can cause a digital system to malfunction.

This shift from personal error to network error has radical implications for accountability. When a human being makes a mistake, there is a clear locus of responsibility. When a system makes a mistake, responsibility diffuses. Who is responsible when a predictive policing algorithm sends officers to the wrong neighborhood? The programmer who wrote the code? The manager who deployed it? The data scientists who trained the model? The institution that chose to use it? The legislators who funded it? The answer is often: no one and everyone. Responsibility evaporates into the complexity of the machinery.

This diffusion of responsibility is one of the most consequential effects of the computer revolution, and it intensifies dramatically with AI. When an AI institution makes a decision — about a loan application, a medical diagnosis, a criminal sentencing recommendation, a hiring decision, and that decision turns out to be wrong, the question of responsibility becomes almost intractable. The AI system did not "decide" in any sense that implies human-like agency. It processed inputs through a mathematical function and produced an output. The "decision" is a statistical artifact, not an act of will. And yet it has real consequences for real people.

The legal and ethical frameworks that govern human decision-making, frameworks built around the concept of individual responsibility, individual intent, and individual accountability — were not designed for frameworks in which decisions are produced by processes that no individual controls, understands, or can be held responsible for. The computer revolution created this problem. AI is making it acute.

There is also the computer's effect on the experience of privacy. Before digital records, privacy was protected primarily by the cost and difficulty of surveillance. Governments, corporations, and individuals could, in principle, monitor each other's activities. But the cost — in time, personnel, and organizational resources — limited the scale and intensity of surveillance. Monitoring was targeted, episodic, and resource-constrained.

Computers eliminated these constraints. Digital systems can monitor continuously, at scale, at near-zero marginal cost. Every transaction, every communication, every movement through a digitally instrumented environment can be recorded, stored, and analyzed. The shift from expensive, targeted surveillance to cheap, universal monitoring is not merely quantitative. It changes the fundamental character of the relationship between the individual and the institutions that govern their life.

In a world of expensive surveillance, privacy is the default. The individual is unmonitored unless specifically targeted. In a world of cheap digital surveillance, monitoring is the default. The individual is continuously observed unless specific measures are taken to prevent it. The burden has shifted from the institution (which must justify and fund surveillance) to the individual (who must actively protect their privacy against a system designed to erode it).

This inversion, from privacy-as-default to surveillance-as-default, is one of the most consequential effects of the computer revolution, and it is largely invisible to most people because it occurred gradually, without any single dramatic moment of change. The shift was not announced. It was not debated. It simply happened, as computerized systems expanded their reach and their capacity, until the experience of being unmonitored became the exception rather than the rule.

That pressure is the real prehistory of algorithmic life.

By the time later arrangements recommend, predict, rank, or generate, the civilization beneath them has already undergone years of training. People have already learned to trust forms, outputs, scores, structured records, and organization-readable identities. The machine does not suddenly invade a fully human world. It enters one already rearranged to accommodate it.

That is why computers belong in this book as more than a bridge chapter. They are the quiet revolution that teaches institutions to feel comfortable outsourcing structure itself.

And once logic has left the human hand, the next step is obvious.

Connection.

If records, messages, and decisions can all become digital, then they can all begin moving instantly.

There is a moment in computing history that reveals civilization's dependency on digital infrastructure with unusual clarity: the Y2K crisis.

As the year 2000 approached, it became apparent that many computer systems, which stored years as two-digit numbers to save memory, would interpret "00" as 1900 rather than 2000. The potential consequences ranged from minor inconveniences (incorrect dates on receipts) to catastrophic failures (banking systems unable to process transactions, power grids unable to maintain operations, air traffic control systems unable to function).

The crisis was averted through an estimated $300 billion global remediation effort, one of the largest coordinated engineering projects in history. Thousands of programmers worked for years to identify and correct the vulnerable code. The effort was largely successful: January 1, 2000, arrived without catastrophic system failures.

The Y2K episode is often remembered as an overreaction, a case of unnecessary panic. This reading misses the point. The panic was appropriate. The reason the catastrophe did not occur was not that the threat was imaginary. It was that the threat was real and was addressed through an enormous, expensive, coordinated effort. Y2K revealed, with unusual clarity, just how deeply civilization had become dependent on computing infrastructure, and how fragile that infrastructure could be. A two-digit storage convention adopted decades earlier for reasons of memory efficiency had become, through the accumulation of dependent systems built on top of it, a potential civilizational vulnerability.

The lesson is directly relevant to AI. The decisions being made now about how AI systems are designed, trained, deployed, and governed will create the infrastructure on which future systems are built. The assumptions embedded in today's AI will become the foundations of tomorrow's civilization, just as the two-digit year convention became the foundation of global computing. And the consequences of those assumptions may not become visible until they have become, like the Y2K bug, too deeply embedded to easily correct.

The internet follows.

There is a phrase that captures the computer's role precisely: "garbage in, garbage out." Meant as a warning about data quality, it contains a deeper truth: the computer rewards precision, structure, and formal consistency while punishing ambiguity and the messy, context-dependent meaning that characterizes most human communication.

Imagine what happens when an institution computerizes its records. Knowledge that existed in informal, distributed form — in the memories of longtime employees, in handwritten notes, in tacit understanding, must be translated into databases, fields, categories, drop-down menus. Information that does not fit must be reformatted or discarded. This process is rarely neutral. A medical record with fields for "diagnosis" and "treatment" but not "patient's emotional state" makes a claim about medically relevant knowledge. A human resources database tracking "performance rating" but not "mentoring contributions" defines institutional values.

Over time, the database shapes the institution's priorities rather than merely recording them. What is measured becomes what is managed. What is managed becomes what matters. What is not captured gradually loses visibility and power.

The philosopher Hubert Dreyfus spent decades arguing this pressure reflected a fundamental misunderstanding of intelligence — that the aspects of human expertise resisting formalization (intuition, judgment, tacit knowledge) are essential features, not deficiencies. Dreyfus lost the institutional argument: computerization proceeded. But his concerns have not been resolved. They have been absorbed into the background.

The computer era introduced a new form of authority: the authority of the output. A number from a computer feels more objective than one from a human, even when based on equally subjective assumptions. "The data shows" has become one of the most powerful authority claims in institutional life. By the time algorithmic governance arrived, the civilization it entered had already been trained to defer to machine outputs. The trust was established. The habit of treating numbers as truth was in place.

And once logic has left the human hand, the next step follows naturally. If records, messages, and decisions can all become digital, then they can all begin moving instantly.

The internet follows.

CHAPTER 12

The Internet: When Distance Died and Boundaries Followed

Distance is usually described as a barrier.

But there was a time when delay protected thought, when unanswered messages remained unanswered without moral consequence, and when not knowing something immediately did not feel like a structure failure. The internet changed that moral atmosphere as much as it changed communication. The internet made distance look like inefficiency.

ARPANET, first operational in 1969, connected a handful of university computers. Tim Berners-Lee's World Wide Web in 1989 made the network accessible through hypertext and browsers. The speed of adoption was unprecedented: radio took thirty-eight years to reach fifty million users; the Web took four years; Facebook took one.

The platformization of social life intensified the transformation. Facebook, Twitter, YouTube, and Instagram provided not merely communication tools but attention environments, continuously optimized to maximize engagement and the generation of behavioral data. The underlying architecture, algorithms determining visibility, notification networks pulling users back, metrics rewarding certain content — was largely invisible to users who experienced the platform as a social space.

The internet also altered the rhythm of obligation. Under networked life, latency becomes interpretable. Silence acquires social meaning. A delayed reply feels intentional. Availability becomes performative.

There is a cost hidden in abundance: the collapse of forgetting. In older systems, life faded because storage was expensive. The internet made memory cheap. Statements linger. Errors persist. Shame becomes searchable. Identity acquires an archive beyond personal control.

That is one reason it spread so quickly and so triumphantly. It collapsed friction in communication, access, publication, commerce, and coordination. Messages that once took days or weeks now take seconds. Information that once required physical proximity now appears instantly. Communities that once depended on geography can gather through networks.

Radio expanded the revolution further. By the 1930s, it had become the first truly mass medium. Franklin Roosevelt's fireside chats demonstrated that a political leader could speak to an entire nation at once, in conversational tone, bypassing institutional intermediaries.[13]

The propagandistic potential was understood immediately. Nazi Germany invested heavily in affordable radio receivers, the *Volksempfänger*, because the regime recognized that broadcast was an environment-shaping instrument.[14] A population tuned to the same frequency, hearing the same voice, could be synchronized at a level of psychological precision that print had never achieved.

Television added the visual dimension. The Kennedy-Nixon debates of 1960 showed that in a visual medium, appearance could overwhelm substance.[15] The Apollo 11 landing in 1969 created the largest shared human experience in history, six hundred million people watching the same event simultaneously.

The advertising industry understood this power with precision. By mid-century, the average American was exposed to thousands of commercial messages per year. Television advertising helped construct a psychology in which desire was not a condition to be occasionally satisfied but a state to be continuously stimulated.

Mass media changes the baseline of awareness. A person inside a mass media environment knows — or believes they know, what is happening everywhere, continuously. This is genuinely valuable. But it normalizes ambient urgency. Crisis becomes background noise. The feeling that something important is happening somewhere becomes a permanent psychological condition.[16]

This transformation feels liberating because it is.

There is a concept in psychology called "continuous partial attention", coined by the technology researcher Linda Stone — that captures the internet's effect on mental life with precision. Continuous partial attention is not the same as multitasking. Multitasking involves dividing attention between multiple tasks for the sake of efficiency. Continuous partial attention involves maintaining a low-level awareness of multiple information streams simultaneously, email, social media, news, messages — not for efficiency but out of a desire to not miss anything important.

The distinction matters because continuous partial attention is not a skill. It is a survival strategy, an adaptation to an environment that continuously signals the possibility of something more interesting, more urgent, or more rewarding arriving at any moment. The person in a state of continuous partial attention is not efficiently managing multiple tasks. They are anxiously scanning multiple channels, afraid that disconnecting from any one of them will cause them to miss something that matters.

This psychological state — always scanning, never fully present, permanently braced for the next interruption, is historically unprecedented. No previous generation of human beings lived in an information environment that demanded constant, low-level vigilance across multiple simultaneous channels. The closest analogy might be a soldier on watch, but soldiers on watch are in recognized states of heightened alertness, and they are relieved after a shift. The internet-connected person is on watch permanently, and no relief arrives.

The consequences for deep work, creative thought, and sustained reflection are difficult to measure precisely but are widely reported. Writers describe increasing difficulty sustaining the concentration required for long-form composition. Academics report that students' ability to read extended texts has declined measurably. Professionals in many fields describe the experience of having their attention fragmented by notifications, emails, and the ambient pull of information streams that never stop.

These reports are not evidence of personal weakness. They are evidence of an environmental change. The human brain did not evolve for continuous partial attention. It evolved for a world in which attention was directed by the person and the task, not continuously solicited by the environment. The internet created an environment that solicits attention constantly — through notifications, alerts, updates, messages, recommendations, and the simple availability of infinite information, and the brain, which is responsive to its environment, adapted. The adaptation is continuous partial attention: a low-grade, permanent alertness that mimics vigilance but actually represents a kind of cognitive surrender to environmental demand.

The result is why the internet chapter precedes the smartphone chapter. The internet created the attention environment. The smartphone made it portable. Together, they produced a condition in which human attention, the most fundamental cognitive resource, the resource that determines what enters consciousness and what does not, became permanently, structurally available to systems designed to capture and monetize it.

It is also destabilizing because distance was never only an inconvenience. It was a protection.

Distance slowed reaction. Distance limited intrusion. Distance protected silence, delay, and local context. Once the internet begins eroding distance, those protections begin eroding as well. Work enters home. News enters every hour. Social comparison expands beyond neighborhood scale. Demands travel farther and faster. The threshold between public and private starts thinning.

There is one more dimension of the internet revolution that this chapter must address before closing: the transformation of the concept of expertise.

Before the internet, expertise was institutionally mediated. If you wanted to know something, you consulted an expert, a doctor, a lawyer, a professor, a mechanic, or you consulted an institution that curated expert knowledge: a library, an encyclopedia, a textbook. The expert's authority derived from training, credentials, and institutional affiliation. The system was imperfect, biased, and often inaccessible. But it provided a rough mechanism for distinguishing between reliable and unreliable knowledge.

The internet disrupted this mechanism by making information universally available regardless of the credentials of its source. A blog post about cancer treatment is, from the browser's perspective, indistinguishable from a peer-reviewed medical journal article. A conspiracy theory about vaccines is delivered through the same interface as a report from the World Health Organization. The internet provides access to everything and authority over nothing.

The result is a phenomenon that Tom Nichols has called "the death of expertise", not the literal disappearance of expert knowledge, but the erosion of the social authority that expertise once carried. In a world where everyone has access to the same information, the expert's advantage, specialized knowledge, trained judgment, institutional credibility — is harder to perceive and easier to dismiss. "I did my own research" has become the rallying cry of a population that has access to unlimited information and diminishing ability to evaluate it.

It is not because people are stupider than they used to be. It is because the information environment has changed faster than the cognitive tools for navigating it. The skills required to evaluate expert claims in a world of curated institutional knowledge are different from the skills required to evaluate claims in a world of uncurated universal information. The first requires trust in institutions. The second requires personal critical thinking at a level that most educational machinerys do not provide and most individuals do not possess.

The internet created a world that demands more critical thinking from ordinary citizens while simultaneously creating conditions that make critical thinking harder: information overload, algorithmic filtering, emotional optimization, and the constant competition for attention that rewards the provocative over the accurate.

The internet does not merely connect. It dissolves boundaries that once seemed natural because they were enforced by friction.

The internet matters here as more than a technological marvel. It changes the baseline of availability. A person reachable only sometimes is different from a person reachable always. A world in which publication is scarce is different from one in which expression is ambient.

The gains are transformative. Access expands. Education spreads. Niche knowledge communities form. Commerce accelerates. Creativity multiplies. But the same forces also generate overload, misinformation, fragmentation, and permanently elevated expectation.

Interruption becomes a condition rather than an exception.

One of the internet's most important psychological effects is that it teaches immediacy. People begin expecting answers faster, transactions faster, responses faster, entertainment faster, resolution faster. The pace of attention changes. Waiting starts feeling more like failure than part of reality.

Here is not simply a preference shift. It is a civilizational adaptation.

The internet also alters publication itself. To publish once required access to institutions or materially costly mechanisms. The web changes that profoundly. This is emancipatory in many respects, but it also means that the distinction between signal and noise becomes less structurally protected by scarcity. Voices multiply. Authority fragments. Trust becomes harder to stabilize.

As with earlier revolutions, what is celebrated first as liberation gradually reveals its more complicated social geometry.

And as with earlier revolutions, the adaptation becomes invisible once it becomes normal. Future generations inherit the internet not as a collapse of historic distance, but as what a connected world obviously is.

The effect is the triumph of normalization. The striking becomes the baseline.

The internet also rearranges the rhythm of obligation. In earlier communication regimes, a letter unanswered for hours or days might mean little. Under networked life, latency itself becomes interpretable. Silence begins to acquire social meaning. A delayed reply can feel intentional. Availability becomes performative. Responsiveness becomes part of identity.

This is a subtle but fundamental shift. The internet does not merely allow contact. It moralizes contact.

A connected person is no longer only someone who can reach others. They are increasingly someone who can be reached, observed, updated, prompted, and expected. The social field becomes more porous. Distance loses not only its physical significance, but its ethical shelter.

That helps explain why the internet feels both emancipatory and exhausting. It widens access while thinning boundaries. It increases voice while reducing insulation. It broadens participation while making disengagement feel less innocent. The person gains unprecedented reach and unprecedented exposure in the same stroke.

This dual movement also affects authority. Under scarcity, institutions filtered more of what counted as publishable knowledge. Under abundance, attention becomes the scarcer resource. This does not eliminate expertise, but it complicates it. Visibility can rise faster than verification. Networks distribute speech faster than judgment. Whole societies become more connected and less epistemically settled at the same time.

There is a dimension of the internet revolution that deserves attention because it captures one of the most important psychological shifts of the digital age: the transformation of the concept of "friends."

Before the internet, the word "friend" had a relatively stable meaning. A friend was a person you knew personally, someone with whom you had a history of direct interaction, shared experience, mutual obligation, and reciprocal care. Friendship was defined by presence, by investment, by the willingness to show up physically when needed.

Social media redefined the word. On Facebook, a "friend" is anyone who has accepted a digital connection request, a coworker you met once, a classmate you have not seen in decades, a stranger who shares your interest in a particular hobby. The average Facebook user has several hundred "friends." The number bears no relationship to Dunbar's number or to any meaningful measure of social intimacy. It is a metric, a count of connections that conflates acquaintance with intimacy, recognition with relationship, and digital proximity with genuine human bond.

This redefinition is not merely semantic. It has psychological consequences. Research on social media use consistently finds that people with more online "friends" do not report greater satisfaction with their social lives. In many cases, they report less. The reason is consistent with what this book has argued: the institution provides the appearance of connection without the substance. Having five hundred "friends" on a platform does not mean having five hundred people who will help you move, listen to your fears, or visit you in the hospital. It means having five hundred people who might see your posts, and whose posts you might see, in a stream of content that algorithms curate for engagement rather than for mutual care.

The internet's redefinition of friendship is one instance of a larger pattern: the systematic redefinition of deep human concepts — community, conversation, connection, privacy, identity, attention, knowledge, by technologies that provide their superficial attributes while stripping their substantive content. A "community" on the internet has members but no mutual obligation. A "conversation" on social media has exchanges but no guaranteed reciprocity. "Connection" on a platform is measured in data but not in care. "Privacy" in the digital age is a setting to be configured, not a condition to be experienced. "Attention" is a commodity to be captured, not a gift to be offered.

In each case, the technology provides a version of the concept that is easier, faster, and more scalable than the original, and that is, for precisely those reasons, thinner. The gain is efficiency and scale. The loss is depth and meaning. And the loss is invisible because the word remains the same. You still have "friends." You still participate in "communities." You still have "conversations." The vocabulary of deep human experience has been preserved. The experiences the vocabulary once described have been transformed.

This linguistic sleight of hand — maintaining the words while changing the realities they describe, is one of the internet's most effective tools of accommodation. It allows the transformation of social life to proceed without linguistic resistance. If the things we lose were renamed — if "friends" became "contacts," "communities" became "audiences," and "conversations" became "exchanges", the losses would be visible. By keeping the old words, the losses disappear into the continuity of language.

This is why the internet is not merely a communications revolution. It is a boundary revolution.

There is a dimension of the internet revolution that is often discussed in terms of politics but that has equally serious implications for psychology: the phenomenon of the filter bubble and its effect on the experience of shared reality.

Before the internet, most people's experience of public information was mediated by institutions that, whatever their biases, served a gatekeeping function. Newspapers had editors. Television had producers. Radio had programmers. These gatekeepers decided what was newsworthy, what was credible, what was appropriate, and what was not. Their decisions were imperfect, biased, and often self-serving. But they produced a shared informational baseline: most people in a given society were exposed to roughly the same major stories, framed in roughly comparable ways.

The internet eliminated this shared baseline. Algorithmic personalization means that two people checking the same platform at the same time may see entirely different content, selected for them by algorithms optimizing for individual engagement rather than collective information. The result is what Eli Pariser called the "filter bubble", an information environment in which each individual is surrounded by content that confirms and reinforces their existing beliefs, preferences, and emotional tendencies.

The political consequences of the filter bubble — polarization, radicalization, the erosion of common ground — have been extensively discussed. Less discussed, but equally important, are the psychological consequences. The experience of shared reality, the feeling that other people inhabit the same world, encounter the same facts, and operate from the same informational foundation — is one of the psychological prerequisites for social cohesion. When people share a common informational baseline, they can disagree about what to do about shared facts. When they no longer share an informational baseline, they disagree about what the facts are — and that kind of disagreement is far more corrosive to social trust.

The internet did not create disagreement. Disagreement is as old as language. But it created a new kind of disagreement: disagreement about the nature of reality itself, fueled by information environments that provide each person with a customized version of the world tailored to reinforce what they already believe. This is not a disagreement that can be resolved through argument, because argument assumes a shared factual foundation from which to reason. When the factual foundation itself is contested, argument becomes impossible and persuasion gives way to tribal assertion.

This erosion of shared reality has implications that extend beyond politics. It affects the capacity for community, for collective problem-solving, for the kind of coordinated action that requires people to agree on what the problem is before they can agree on what to do about it. Climate change, pandemic response, economic policy, educational reform, every complex social challenge requires a shared understanding of the relevant facts as a prerequisite for productive disagreement about solutions. The filter bubble threatens that prerequisite.

The internet also accelerated a phenomenon that has implications for every subsequent chapter: the collapse of the distinction between production and consumption.

Before the internet, most people were primarily consumers of media — they watched television, read newspapers, listened to radio. Production was concentrated in the hands of professionals working within institutions. The internet blurred this distinction by giving every connected individual the tools to publish, broadcast, and distribute content to a global audience. This democratization of production is, in many respects, genuinely liberating. It has given voice to marginalized communities, enabled new forms of creative expression, and held powerful institutions accountable in ways that traditional media could not.

But it has also created an information environment in which the volume of content vastly exceeds the capacity of any human mind to evaluate it. In a world of professional gatekeeping, the volume of published content was limited by the capacity of institutions to produce it. In a world of universal publication, the volume is limited only by the number of people with internet access and something to say. The result is an information environment that is simultaneously richer and more chaotic than any previous one, offering more perspectives, more data, more voices, and more possibilities than ever before, while also offering more noise, more misinformation, more manipulation, and more cognitive overload.

The human mind was not designed for this environment. It was designed for a world in which the volume of information was limited, the sources were familiar, and the evaluation of claims could draw on direct personal knowledge and community reputation. The internet created a world in which none of these conditions hold, and the cognitive and emotional consequences of that mismatch are still being discovered.

It changes what counts as private enough, delayed enough, distant enough, and separate enough to preserve older habits of thought. The person no longer encounters the world only in chunks defined by place and time. The world becomes increasingly ambient, searchable, and pushable.

There is another cost hidden in this abundance: collapse of forgetting. In older networks, much of life faded because storage, retrieval, and circulation were expensive. The internet lowers those costs dramatically. Statements linger. Images recur. Errors persist. Shame can become searchable. Identity acquires an archive.

This changes the psychology of self-presentation. To speak online is not merely to express. It is often to deposit a version of oneself into systems that may retrieve it later under different conditions, different audiences, and different stakes. The internet does not only widen speech. It lengthens consequence.

The result is a new kind of vigilance. A person learns not only to communicate, but to manage trace. Memory is no longer only human and local. It becomes infrastructural.

That infrastructural memory will matter enormously in the next chapter, where the network becomes small enough, portable enough, and constant enough to stop feeling like a place one goes and start feeling like a layer of consciousness itself.

Yet even the internet is not intimate enough. For all its reach, it initially still often requires terminals, desks, intentional logins, periods of connection and disconnection.

The next step removes even that residual distance.

There is a small, telling phenomenon that captures the smartphone's colonization of the nervous system: phantom vibrations.

Studies conducted since the late 2000s have found that a significant majority of smartphone users report occasionally feeling their phone vibrate when it has not, in fact, vibrated. The sensation is so common that researchers have given it a name: "phantom vibration syndrome." The experience is not hallucination in any clinical sense. It is a perceptual bias, a recalibration of the sensory threshold at which the brain interprets ambiguous tactile stimulation (a muscle twitch, a clothing shift, a slight pressure change) as a phone notification.

The phenomenon is minor in itself. No one suffers from phantom vibrations. But it is diagnostically significant for the argument of this book. It demonstrates that the smartphone has altered the brain's sensory processing, not through any deliberate design but through the sheer frequency of repetition. The brain, having received thousands of real vibration signals over months and years, has lowered its detection threshold to the point where it generates false positives. The phone has, in a literal neurological sense, rewired the body's sensory apparatus.

If a device can recalibrate tactile perception through nothing more than repeated use, what other perceptual and cognitive recalibrations might it be producing? The smartphone user who feels phantom vibrations is experiencing the most visible symptom of a much broader process: the progressive adaptation of the nervous system to a technological environment that it was not evolved to inhabit. The adaptation is automatic, unconscious, and — like every adaptation described in this book — invisible to the person undergoing it.

The person who feels a phantom vibration does not think: "My nervous system has been recalibrated by repeated technological stimulation." They think: "I thought my phone buzzed." The extraordinary has, once again, become ordinary. The recalibration has disappeared into the background of daily experience.

The network enters the pocket.

Take a person in 1985 — an office worker in Minneapolis. Their morning begins with a local newspaper and coffee. They drive to work with local radio. At the office, they see the same colleagues daily. Phone calls connect them to a few cities. Mail arrives once daily. Their social world is bounded by geography, time, and physical communication costs. They are located — rooted, surrounded by knowable community at a scale matching human cognitive capacity.

Now the same person in 2015. Before rising, they check email from three time zones. They scroll a news feed mixing local events with international crises, celebrity gossip, and algorithmic outrage. Video calls span four continents. Slack messages outpace reading. Social media notifications pull at attention all day. In the evening they stream Korean television, argue politics with strangers, and fall asleep to algorithmically recommended videos.

This person is no longer located. They are dispersed, stretched across simultaneous social contexts, time zones, and information streams, with no clear boundary between here and there, work and home.

The psychological consequences are becoming visible. Heavy internet users report higher social comparison, information overload, decision fatigue, and a persistent sense of being always available, always reachable, always accountable. These are structural consequences of an information environment with no edges, no closing time, and no natural boundaries.

Yet even the internet retained a residual distance. It required terminals, desks, intentional logins. There remained moments when a person was offline, unreachable, alone with their own thoughts.

The next revolution would eliminate even that.

The network enters the pocket.

CHAPTER 13

The Pocket God: When Attention Became Extractable

The smartphone is not merely a tool. It is an environment carried in the hand.

Here lies why it became so consequential so quickly. The smartphone condenses communication, entertainment, navigation, memory, social performance, commerce, surveillance, work, and identity into a single portable object that remains within reach almost all the time. Previous media demanded attendance. The smartphone demands residency.

Once that happens, attention becomes continuously available.

The result is one of the most important transformations in modern life. Human attention, once scattered across idle moments, physical surroundings, and inner continuity, becomes increasingly measurable, targetable, and harvestable. The idle interval becomes monetizable. The glance becomes data. The pause becomes opportunity.

The smartphone is central to this book because it does not merely add convenience. It installs extraction inside everyday consciousness.

Think about the morning ritual of contemporary life. Before the body has fully entered the day, the hand reaches. The screen lights. Time, weather, messages, demands, alerts, approvals, market signals, social updates, and ambient urgency flood inward before unstructured thought has had much chance to form. This is not a universal ritual, but it is common enough to be historically revealing. The device no longer waits to be used. It waits to resume the user.

The gains are undeniable. Maps, emergency contact, instant communication, portable knowledge, on-demand access, social connection, digital commerce, photography, reminders, and tools of every kind now sit in one place. The device is genuinely useful in a way many criticisms fail to acknowledge.

But usefulness is precisely how dependency deepens.

The smartphone changes more than behavior. It changes thresholds. Silence feels heavier. Waiting feels less tolerable. Solitude becomes easier to escape and therefore harder to practice. External validation becomes more ambient through metrics, notifications, and social visibility. The device does not force these changes single-handedly, but it normalizes them.

The person no longer merely uses the phone. The phone becomes part of the structure through which the person experiences the day.

This matters because attention is not just another resource. Attention is the gating function of consciousness. To shape attention repeatedly is to shape mood, priority, memory, and desire. The more human beings live within systems optimized to reclaim their attention, the more their inner life begins sharing territory with market design.

The smartphone is therefore not just a communications breakthrough. It is a portable regime of incentives.

Alerts produce anticipation. Metrics produce comparison. Infinite scroll produces temporal blur. Recommendation engines reduce the distance between impulse and consumption. Cameras and feeds turn private experience into potential performance. Identity grows more externalized. The self is increasingly tracked in impressions, responses, visibility, and engagement.

This does not erase individuality. It changes the terrain on which individuality is performed.

A pocket device becomes a pocket authority. It tells us where to go, whom to answer, what to notice, what to fear, what to buy, what to miss, and what we may be missing right now.

The astonishing thing is not that people accepted this. The astonishing thing is how quickly it ceased to feel astonishing.

That is normalization at its most intimate.

A person born into the smartphone age does not experience permanent networked access as a rupture. It appears as the shape of the world. The historical novelty collapses into baseline expectation. A device capable of mapping, tracking, storing, summoning, publishing, measuring, and interrupting becomes so intimate that its absence can feel more jarring than its presence.

The smartphone normalized permanent partial occupation of the mind.

This sentence should not be mistaken for exaggeration. Permanent occupation does not mean total control. It means that a substantial portion of consciousness remains available for reclamation by designed systems. The body may be physically elsewhere. The mind remains partially hookable.

This is also why the smartphone chapter belongs close to the AI chapter. The smartphone trains the conditions under which later cognitive institutions become acceptable. It accustoms people to carrying systems that know, remind, nudge, suggest, compare, and mediate. Once such mediation becomes ordinary, more advanced forms of delegation feel less like rupture and more like improvement.

The pocket device teaches the mind to welcome assistance and tolerate interruption under the same icon.

The result is not only social or commercial. It is existentially rhythmic. The day begins to fracture into countless invitations to re-enter the device's logic. The person does not simply check a phone; the person repeatedly crosses a threshold between embodied life and apparatus-shaped attention. After enough repetitions, the crossing itself becomes invisible.

That invisibility is where the revolution fully succeeds.

There is a final observation about the smartphone that connects it to every chapter that follows: the device's role as a training apparatus for AI dependence.

Consider the psychological habits the smartphone installs. It teaches you to consult a device before making decisions — which restaurant, which route, which product, which movie. It teaches you to outsource memory, why remember a phone number, an address, a fact, when the device remembers for you? It teaches you to expect instant answers, why sit with uncertainty when the answer is a search away? It teaches you to seek external validation, why trust your own judgment about a photograph, a meal, an experience, when metrics of social approval are immediately available?

Each of these habits is, individually, reasonable. Each represents a genuine convenience. But collectively, they constitute a psychological training program, a systematic habituation to the experience of cognitive dependence on an external system. The smartphone user who has spent a decade consulting their device for directions, for facts, for social validation, and for emotional regulation has been trained, through daily repetition, to treat the device as a cognitive extension of themselves.

When AI enters this person's life, not as a separate technology but as an upgrade to the device they already carry and the habits they already possess, the adoption will feel not like a revolution but like an improvement. The AI assistant is simply a better version of what the smartphone already provides: better answers, better recommendations, better anticipation of needs. The person does not need to be convinced to trust AI. The smartphone has already done that work.

This is why the smartphone chapter comes immediately before the AI chapter in this book's architecture. The smartphone is not merely a communication device. It is a psychological preparation for the next phase of human-machine integration. It trains the habits, expectations, and tolerances that make AI adoption feel natural rather than alien. It is the gateway drug of cognitive dependence.

The deeper cost is not only distraction. It is erosion of psychic edges. A mind that is continually reachable becomes different from a mind that can disappear into itself without contest. The smartphone does not make depth impossible, but it makes depth more interruptible. It changes how often thought completes its own arc before another claim enters.

That matters because inner continuity is one of the least defended human resources. Few people notice it directly until it has already thinned. They notice restlessness. Reflex checking. Anxiety at silence. The feeling that an unfilled moment must mean something is missing. These sensations are not random. They are the psychological weather of an attention economy that has become intimate enough to live in the hand.

The device also alters memory in a subtle way. When photographs, messages, maps, reminders, contacts, and search live externally and access is immediate, the burden of remembering changes. This is convenient and often beneficial. But convenience can conceal cognitive migration. The person increasingly remembers with the device rather than simply through the mind. The self becomes distributed across systems that feel personal because they are always near.

There is another shift here that matters for the book's darker undertone: anticipation becomes monetized. Before the phone is checked, there is the possibility that something has happened. A message. A change. A missed opportunity. An approval. A threat. A piece of proof that one still matters to the day. This anticipatory charge is small in any individual moment, but civilizationally large when repeated enough. The device becomes not only a tool of access, but a machine for training expectation.

And expectation, once trained, becomes demand.

The point is why the smartphone is more than a digital Swiss Army knife. It is the everyday tutor of impatience. It teaches the nervous system that response should be quick, friction should be minimal, boredom should be avoidable, and quiet should be interruptible. Each lesson is small. Their aggregate effect is enormous.

This also helps explain the peculiar emotional ambiguity of the device. People resent it and reach for it. They feel trapped by it and protected by it. It is resented as intrusion and relied upon as reassurance. That combination is one of the clearest signs that a structure has become infrastructural. It is no longer merely used. It has become part of how stability is felt.

The smartphone's effect on the experience of boredom deserves special attention, because boredom, far from being a trivial complaint — may be one of the most psychologically important experiences that the smartphone is eliminating.

Boredom is uncomfortable. It is the experience of unfilled time, of consciousness with nothing to attach to, of the mind thrown back on its own resources. Most people, when asked, will describe boredom as something to be avoided. Entire industries — entertainment, social media, mobile gaming, are built on the proposition that boredom is a problem and their product is the solution.

But psychologists who study boredom have found something more nuanced. Boredom, they argue, is not merely the absence of stimulation. It is a signal — a psychological prompt that motivates the individual to seek new goals, new activities, new directions. Boredom is the mind's way of saying "what you are currently doing is not meaningful enough. Find something that is." In this sense, boredom is not an enemy of creativity and purpose. It is a precondition for them. Many of the most important human experiences, the decision to learn a new skill, the impulse to start a creative project, the recognition that a relationship or a career needs to change, begin with boredom. They begin with the uncomfortable awareness that the current state of affairs is insufficient.

The smartphone eliminates this awareness by providing an instant, effortless escape from boredom. The moment consciousness begins to experience unfilled time, the hand reaches for the device, and the screen provides a flood of stimulation that suppresses the boredom signal before it can complete its psychological function. The person is no longer bored, but they have also lost the motivational prompt that boredom was designed to provide.

Over time, this creates a subtle but potentially consequential change in the inner life of the individual. A person who has never sat with boredom long enough to feel its motivational pressure is a person who may never experience the internal push toward new goals, new activities, and new forms of meaning. They are a person whose sense of purpose is increasingly supplied from outside, by feeds, by recommendations, by the next notification, rather than generated from within by the mind's own restless search for significance.

Research by Timothy Wilson and colleagues at the University of Virginia found that many people would rather administer mild electric shocks to themselves than sit alone with their thoughts for fifteen minutes. This finding, widely reported as evidence of modern attention deficit, may actually reveal something deeper: a population that has lost the ability to tolerate the experience of consciousness without external stimulation — not because their brains have changed, but because the environment has eliminated the need to develop that tolerance.

The smartphone did not create the human desire to avoid boredom. That desire is ancient. What the smartphone did was satisfy that desire instantly, continuously, and without limit, and in doing so, it may have disrupted a psychological mechanism that had been functioning, uncomfortably but usefully, for the entire history of the species.

There is also the smartphone's relationship to sleep. Studies consistently find that smartphone use before bed delays sleep onset, reduces sleep quality, and alters circadian rhythms — primarily through the blue light emitted by screens, which suppresses melatonin production, but also through the cognitive arousal produced by social media, email, news, and other stimulating content. The World Health Organization and numerous national health agencies have identified sleep deprivation as a significant public health concern in developed countries, with consequences ranging from impaired cognitive function to increased risk of chronic disease.

The smartphone is not the sole cause of the modern sleep crisis. But it is a significant contributor — and its contribution illustrates the book's broader argument about invisible costs. The person who scrolls their phone in bed at midnight does not experience themselves as participating in a public health crisis. They experience themselves as checking one more thing before sleep. The cost is invisible because it is statistical — it manifests as a slightly worse night's sleep, slightly impaired concentration the next day, slightly elevated stress, slightly increased risk of health problems over years. No single night's scrolling is catastrophic. The aggregate effect, across millions of people and thousands of nights, is civilizationally significant.

Here lies the structure of every cost this book has documented: small enough to ignore individually, large enough to matter collectively, and invisible enough to escape notice until the damage is done.

There is a final dimension of the smartphone revolution that deserves attention because it captures, in miniature, the entire argument of this book: the disappearance of the transition between connected and unconnected states.

Before the smartphone, being "online" was a discrete state. You sat down at a computer. You logged on. You checked email, browsed the web, communicated with others through the network. Then you logged off. You stood up. You walked away from the computer. And in the interval between sessions, you were offline — unconnected, unreachable, alone with your own thoughts and the physical world around you.

The smartphone eliminated that transition. There is no longer a moment of logging on, because you are never logged off. There is no longer a moment of sitting down at the computer, because the computer is in your pocket. There is no longer a boundary between the connected state and the unconnected state, because the connected state has become continuous.

This elimination of the transition is, in a precise sense, the elimination of a boundary — and boundaries, as this book has argued throughout, are not merely inconveniences. They are structures that shape experience. The boundary between work and home shapes the experience of both. The boundary between day and night shapes the experience of both. The boundary between online and offline shapes the experience of both. Remove the boundary, and the experiences on either side of it change, not because they are forbidden, but because they are no longer differentiated.

A person who is always online does not have the same relationship with being online as a person who is sometimes online. The permanent state does not merely extend the duration of connectivity. It changes its character. Being online becomes not an activity but a condition, not something you do but something you are. And conditions, unlike activities, are invisible. You do not notice the air you breathe. You do not notice the gravity that holds you down. You do not notice the connectivity that surrounds you — until it fails, and the sudden absence of what you had stopped noticing produces a disorientation that reveals, with startling clarity, how thoroughly the condition had become part of your experience of reality.

This is the smartphone's deepest achievement: the elimination of the last boundary between the human and the network. Previous technologies approached this boundary — the laptop made connectivity portable, the BlackBerry made email continuous, but the smartphone eliminated it. There is no longer a gap. The human and the network are, for practical purposes, continuously joined.

That joining is the culmination of a process that began with fire. Fire joined the human to the tool. Language joined the human to the symbol. Agriculture joined the human to the schedule. The city joined the human to the order. Writing joined the human to the record. Money joined the human to the abstraction. Industry joined the human to the clock. Media joined the human to the broadcast. The computer joined the human to the process. The internet joined the human to the network.

The smartphone made the joining permanent.

And AI may make it indistinguishable from thought itself.

The effect is the trajectory this book has traced. Not a single event but a progressive deepening, a narrowing of the distance between the human being and the systems that shape their experience, until the distance disappears entirely and the machinerys become, for all practical purposes, part of the self.

Here lies a powerful precondition for the next revolution.

Once attention is extractable, thought itself becomes the next frontier.

There is one more dimension of the smartphone revolution that connects it to the book's deepest concerns: the device's effect on the experience of presence.

Presence, the experience of being fully located in the current moment, in the current place, with the current people, is something that most people take for granted, or rather, took for granted. Before the smartphone, presence was the default mode of human experience. You were where you were. You were with whom you were with. Your attention was directed by the situation you inhabited. Distraction existed, of course — daydreaming, boredom, the wandering mind — but the distractions were internal. They came from within the self, and they could be managed through the same internal resources that generated them.

The smartphone externalized distraction. It made distraction not merely possible but continuously available, continuously tempting, and — through the design logic of notification systems, variable reward schedules, and social obligation — continuously demanding. The result is a new default mode: partial presence. The body is in one place. The mind is simultaneously monitoring another — the stream of notifications, the possibility of new messages, the ambient awareness of what might be happening elsewhere.

This partial presence has consequences for every form of human interaction. A conversation in which both participants are partially present is a different kind of conversation than one in which both are fully present. Research on phone-mediated social interaction consistently finds that the mere visible presence of a smartphone on a table during a conversation — even if it is face-down and silent, reduces the depth and quality of the exchange. Participants report feeling less connected, less empathetic, and less satisfied than in conversations where no phone is visible. The phone does not need to ring or buzz to exert its influence. Its mere presence signals the possibility of interruption, and that signal is enough to prevent full engagement.

Picture what this means for the deepest forms of human connection: the heart-to-heart conversation, the intimate confession, the shared silence between people who know each other well. These experiences depend on presence — on the mutual understanding that, for this moment, nothing else exists except the two people and what passes between them. The smartphone does not forbid these experiences. But it places them in competition with an always-available alternative, and the alternative, by the logic of novelty and variable reward, often wins.

Parents report checking their phones while playing with their children. Partners report scrolling feeds during conversations. Friends report the ambient frustration of talking to someone whose eyes keep drifting to a screen. In each case, the complaint is the same: the person is physically present but psychologically elsewhere. The body is in the room. The attention has been captured by a device in the pocket.

What emerges is not a moral failing. It is a design outcome. Smartphones are engineered to recapture attention. Notification systems are designed to interrupt. Variable reward schedules, the same psychological mechanism that makes slot machines addictive, are embedded in the design of social media feeds. The user is not weak for being distracted. The user is responding rationally to a system that has been optimized, through millions of dollars of research and design, to reclaim their attention at every opportunity.

The question is not whether individual users can resist the pull of their devices through willpower and discipline. Some can, sometimes. The question is what happens to a civilization in which the default mode of consciousness — the ordinary, everyday, background state of the human mind, shifts from presence to partial presence. What happens to intimacy, to creativity, to reflection, to the quality of democratic citizenship, to the experience of being alive, when the mind is never fully where the body is?

That question does not have a definitive answer yet. The experiment is still running. But the preliminary data, the rising rates of anxiety, depression, and loneliness in smartphone-saturated populations, the declining capacity for sustained attention, the reported sense of chronic distraction and dissatisfaction, suggest that the costs of partial presence may be substantial.

The smartphone did not merely change what people do with their time. It changed the quality of their presence in their own lives.

The machine will not only interrupt.

It will increasingly assist, mimic, and compete.

PART V

The Crisis of Human Uniqueness

The smartphone's effect on childhood deserves particular attention because it reveals the pattern of this book in its most intimate and consequential form. A child born after 2010, a member of what Jean Twenge has called "iGen", grows up in a world where smartphones are not novelties but furniture. The device is present in the home from birth. It is used by parents, siblings, teachers, and peers. It mediates entertainment, communication, education, social interaction, and, increasingly — emotional regulation.

Consider what this means developmentally. A child's brain is shaped by its environment during critical periods of development. The neural pathways that govern attention, impulse control, social cognition, emotional regulation, and executive function are all built through repeated experience during childhood and adolescence. A child who grows up in a smartphone-saturated environment develops neural pathways adapted to that environment — pathways optimized for rapid task-switching, short-duration attention, externally driven stimulation, and constant social comparison.

What matters is not speculation. It is consistent with decades of neuroscience research on environmental influence on brain development. The specific effects of smartphone-saturated childhoods are still being studied, and the evidence is not yet conclusive on all points. But the direction of the findings is consistent and concerning: increased rates of anxiety, depression, sleep disruption, and attentional difficulty among adolescents correlate with the period of smartphone saturation, and the correlation is particularly strong among girls and among heavy users.

Jonathan Haidt and others have argued that the combination of smartphone-based social media and the decline of unsupervised childhood play has produced what amounts to a rewiring of adolescent social development, replacing the slow, embodied, face-to-face social learning that characterized previous generations with a fast, disembodied, performance-based social environment that amplifies comparison, reduces resilience, and creates a permanent audience for every adolescent mistake.

Whether or not one accepts Haidt's specific framing, the underlying concern is difficult to dismiss: the smartphone is not merely a tool that children use. It is an environment that children grow up inside. And environments shape organisms. The child who develops inside the smartphone environment is not the same organism as the child who developed inside the playground environment, any more than the fire-keeping hominid was the same organism as the fire-fearing one.

The difference is speed. Fire reshaped human biology over hundreds of thousands of years. The smartphone is reshaping human psychology over a single generation. There is no evolutionary precedent for an environmental shift of this magnitude occurring this quickly. We are, in effect, running a civilization-scale experiment on the developing brains of an entire generation, and we are running it without controls, without consent, and largely without awareness.

That last point is the one most relevant to this book's thesis. Most parents who give their children smartphones do not experience themselves as participants in an experiment. They experience themselves as doing what everyone else is doing, which is to say, they experience themselves as normal. The child who grows up with a smartphone does not experience themselves as a subject of an unprecedented developmental intervention. They experience themselves as a kid with a phone.

That is normalization in its purest form. The experiment disappears into the background of ordinary life. The extraordinary becomes invisible. And the organism is reshaped without ever recognizing that reshaping is underway.

If earlier revolutions changed labor, time, memory, and attention, the next revolutions move closer to the core of human self-understanding. Intelligence, authenticity, intimacy, and agency begin to look less securely human than they once did.

What is at stake in this part is not simply utility, but self-conception. Human beings begin confronting systems that do not merely support activity. They begin confronting orders that reflect, rival, and reshape meanings once reserved for persons.

CHAPTER 14
AI: When Thought Met Competition

For most of human history, tools extended muscle, reach, memory, transport, or communication.

What makes AI historically jarring is not simply that it performs tasks. It enters a zone many humans treated as existentially protected: language, judgment, synthesis, and the outward performance of thought. Artificial intelligence is different because it enters territory humans long treated as evidence of mental uniqueness.

It writes. It summarizes. It translates. It classifies. It predicts. It generates. It recommends. It reasons in ways that feel, at minimum, adjacent to human cognitive behavior.

The result is why AI unsettles so deeply. The disruption is not only economic. It is also existential.

A machine lifting more than a person may impress. A machine drafting, analyzing, or solving in language begins pressing on a different nerve. It forces the question humans are least comfortable asking: what if thought is not as exclusively ours as we believed?

That question does not require consciousness to become destabilizing. Even arrangements that are not sentient can alter human self-understanding if they perform enough tasks once treated as uniquely cognitive. Human beings do not reserve their existential anxiety only for metaphysical threats. Functional comparison is enough.

Here is what makes AI a threshold chapter in the history of normalization. It is not simply another automation wave. It is a challenge to mental exclusivity.

The gains are exceptional. Knowledge work accelerates. Drafting, coding, brainstorming, pattern recognition, search, support, translation, and analysis become faster and more widely accessible. People with less experience can perform closer to expert levels in some domains.

Organizations can process more with fewer frictions. For many users, AI appears first not as domination but as relief.

Again, relief is part of the seduction.

AI removes effort in places where effort once anchored identity. It makes difficult things easier. It reduces cognitive burden. It gives ordinary users access to synthetic competence on demand.

The point is why it will be absorbed so fast.

But every tool that removes strain also risks eroding the significance of what strain once built. If writing, coding, summarizing, planning, and ideation become increasingly delegated, what happens to the human experience of mastery? What happens to authorship when production is co-generated? What happens to confidence when comparison is no longer only among humans?

The real shock of AI may not be replacement. It may be relativization.

A person may remain employed and still feel displaced in a deeper sense. A person may still create and still wonder whether creation means the same thing when machine assistance can simulate so much of the process. A person may still think and yet increasingly think with systems that reduce the felt necessity of independent synthesis. This is not straightforward loss. It is reclassification.

Human beings may continue doing many valuable things. But they may do them in the presence of infrastructures that make old definitions of competence feel less stable. This can produce liberation, insecurity, dependency, denial, and exhilaration all at once.

AI is therefore not merely an economic disruption. It is a psychological mirror placed before the species.

And if its technical trajectory continues, the mirror may become increasingly uncomfortable.

Part of AI's normalization power comes from the speed with which it migrates from spectacle to workflow. An astonishing demo becomes a workplace feature. A public debate becomes a procurement decision. A cultural panic becomes a productivity discussion. This sequence is familiar now, but it is still historically unusual. Few revolutions so quickly enter the practical language of ordinary work while simultaneously challenging deep assumptions about what human beings are for.

And so the emotional response to AI is often mixed in ways earlier tools did not provoke. People feel empowered and diminished, impressed and suspicious, efficient and strangely hollow. The machine does not merely save time. It alters the meaning of using time well.

There is also a status dimension. For centuries, many forms of prestige rested partly on the scarcity of cognitive performance: the ability to summarize well, reason clearly, draft persuasively, synthesize broadly, or produce quickly under pressure. AI does not destroy these qualities, but it redistributes them. What was once a signal of individual capability can become, at least partly, a signal of access to machine assistance.

There is one more dimension of AI that connects it to the book's deepest argument: the question of what happens to the concept of originality.

In the pre-AI world, originality was a meaningful concept. A piece of writing, a work of art, a musical composition, a scientific theory, each could be evaluated, at least roughly, in terms of its novelty. Was this idea new? Had this combination of elements been attempted before? Did this work add something to the conversation that was not there previously? The answers to these questions were imperfect, but the questions themselves were meaningful because the production of intellectual and creative work required human effort, and human effort produces variation rooted in individual experience, perspective, and cognitive style.

AI complicates this concept fundamentally. A language model trained on the corpus of human writing can generate text that is, in a statistical sense, novel — it combines words and ideas in sequences that do not appear in its training data. But the novelty is not rooted in experience, perspective, or individual cognitive style. It is rooted in mathematical operations over probability distributions. The output is new in the sense that it has not been produced before. It is not new in the sense that it reflects a new way of seeing, thinking, or feeling.

This distinction may seem philosophical. It has practical consequences. If the concept of originality becomes unmeanable, if the sheer volume of AI-generated content makes it impossible to distinguish between genuinely novel human insight and statistically novel machine output — then the social structures that depend on originality as a criterion of value may collapse. Academic publishing depends on originality. Patent law depends on it. Artistic reputation depends on it. Educational assessment depends on it. Each of these systems assumes that originality is detectable and attributable to a specific human mind. AI undermines both assumptions.

The deeper question is whether originality, as a human value, can survive the automated production of novelty. The answer may depend on whether civilization can maintain the distinction between two kinds of newness: the newness of recombination (which AI excels at) and the newness of vision (which, at least for now, remains a human capacity). The first produces outputs that are statistically unprecedented. The second produces outputs that reflect a way of seeing the world that only a particular mind, shaped by a particular life, could have generated.

If that distinction is preserved — if civilization continues to value the newness of vision over the newness of recombination, then AI may enhance rather than replace human creativity, by handling the recombinative labor and freeing humans for the visionary work. If the distinction is lost — if the sheer volume and quality of AI-generated content makes the concept of individual creative vision seem quaint or unnecessary, then something important about human culture will have been traded for efficiency.

And the trade, as always, will have been made so gradually, so practically, and so beneficially that it will barely be noticed.

That redistribution is quietly destabilizing.

There is a dimension of AI that most discussions overlook because it is uncomfortable: the question of what happens to human motivation in a world where machines can do most cognitive tasks better, faster, and cheaper than people.

Consider the arc of a typical professional career in the pre-AI world. A young lawyer spends years learning to research, analyze, draft, and argue. Each year, their skills improve. Each improvement brings greater responsibility, higher compensation, and — crucially, a sense of growing competence. The experience of becoming good at something difficult is one of the primary sources of meaning in professional life. It is not merely about the money. It is about the feeling that one's efforts are producing measurable improvement in one's capabilities, that one is, through sustained effort, becoming more capable than one was before.

Now consider the same career in an AI-augmented world. The young lawyer has access to systems that can research, analyze, and draft at a level that previously required years of training. The systems do not replace the lawyer, judgment, strategy, and client relationship still matter. But they dramatically reduce the proportion of the work that requires the kind of skill-building effort that previously constituted the core of professional development. The young lawyer may be more productive. But they may also be less engaged — because the difficult, frustrating, ego-bruising process of learning to do hard things has been partially automated.

This is not a trivial concern. Psychologists who study motivation, most notably Mihaly Csikszentmihalyi, in his research on "flow", have found that the experience of deep engagement depends on a specific balance between challenge and skill. When the challenge exceeds the skill, the result is anxiety. When the skill exceeds the challenge, the result is boredom. When challenge and skill are matched, the result is flow, a state of absorbed, energized, intrinsically motivated engagement that is associated with both high performance and high well-being.

AI threatens to disrupt this balance across a wide range of professional and creative activities. By handling the routine cognitive challenges that previously occupied much of a professional's day, AI may push many workers into a zone where their skills exceed the remaining challenges, the zone of boredom, underemployment, and diminished meaning. The worker is still employed. They are still compensated. But the part of the work that made them feel competent, engaged, and professionally alive has been delegated to a machine.

This is one of the more insidious potential consequences of AI, because it is not visible in productivity statistics. A law firm whose associates use AI to draft contracts may show increased output and reduced costs. The associates themselves may report satisfaction with the efficiency gains. But over time, the cumulative effect of delegating skill-building tasks to machines may be a generation of professionals who are less deeply trained, less intrinsically motivated, and less capable of the kind of independent judgment that comes only from having done the hard work themselves.

The parallel with earlier revolutions is instructive. The assembly line increased manufacturing productivity but deskilled individual workers, replacing craftsmen who understood the entire production process with operators who performed a single repetitive task. The efficiency gains were real. The cost, in terms of worker autonomy, job satisfaction, and the preservation of craft knowledge — was equally real but harder to measure. AI may be performing a similar trade in the cognitive domain: increasing aggregate productivity while quietly eroding the individual experience of meaningful work.

There is also the question of what AI does to the experience of creativity, perhaps the most valued and most vulnerable of human cognitive capacities.

When a person creates something, a piece of writing, a piece of music, a design, a solution to a problem, the experience is not merely one of producing an output. It is an experience of self-expression, of bringing something into existence that did not exist before, of imposing one's own perspective and aesthetic on the world. Creativity, in this sense, is not just a cognitive function. It is a mode of being, a way of relating to the world that is deeply connected to identity, meaning, and the sense of being alive.

AI can produce outputs that, by external criteria, are indistinguishable from creative human work. It can write poems, compose music, generate images, and propose solutions that would, if attributed to a human creator, be praised for their creativity. But the AI does not experience the creation. It does not struggle, revise, doubt, or feel the satisfaction of having brought something new into being. The output exists, but the experience that made the output meaningful — the human experience of creating — is absent.

The question is whether a civilization that increasingly delegates creative work to machines will also lose the experience of creation, the struggle, the frustration, the satisfaction, the meaning — that has been one of the most valued dimensions of human life. The outputs may persist. The experience may fade. And if the experience fades, something important about what it means to be human fades with it.

The tool helps. The user benefits. But the boundary around personal accomplishment becomes less secure.

There is a dimension of the AI revolution that most discussions avoid because it touches on something that modern societies find deeply uncomfortable: the question of what happens to social status and self-worth when cognitive capability is no longer a reliable marker of human value.

For most of the last century, the social hierarchy of developed nations has been organized, to a significant degree, around cognitive capability. Intelligence — as measured by educational achievement, professional competence, verbal fluency, and the ability to process complex information — has been the primary currency of social status in knowledge economies. The most valued, most compensated, most socially respected individuals in modern societies tend to be those who demonstrate the highest cognitive capabilities: doctors, lawyers, engineers, scientists, executives, professors.

AI threatens this hierarchy not by eliminating cognitive work but by democratizing it. When an AI system can produce competent legal analysis, medical diagnosis, financial modeling, strategic planning, and creative writing, the scarcity that made these skills valuable diminishes. The lawyer's expertise is still useful — but it is less scarce when AI can approximate it. The doctor's diagnostic skill is still important, but it is less distinguishing when AI can match it in certain domains. The writer's craft is still valuable, but it is less rare when AI can produce serviceable prose on demand.

The economic consequences of this shift are widely discussed. The psychological consequences are less so, but they may be equally significant. For millions of people whose sense of identity and self-worth is tied to their cognitive capabilities — to the feeling that they are smart, competent, and valued for what they know and can do — the arrival of arrangements that approximate those capabilities at near-zero cost represents not merely an economic disruption but an existential one.

The result is not an argument against AI. It is an argument for taking seriously the psychological dimensions of a technological transition that is often discussed in purely economic terms. The question is not only "how will we distribute the economic gains of AI?", though that question is important. The question is also "how will human beings maintain a sense of purpose, dignity, and self-worth in a world where their most valued cognitive capabilities can be approximated by machines?"

Previous technological revolutions encountered similar questions. The mechanization of physical labor threatened the status and self-worth of craftsmen whose skills were rendered obsolete by machines. The transition was painful, prolonged, and socially disruptive, involving decades of resistance, reform, and eventual adaptation. The result was a reorientation of social value from physical skill to cognitive skill: the knowledge economy emerged partly as a response to the automation of manual labor.

AI may require a similar reorientation, but it is not obvious what the new basis of human value will be. If neither physical skill (automated by machines) nor cognitive skill (approximated by AI) provides a reliable foundation for human status and self-worth, then the civilization must find a new one. Emotional intelligence? Ethical judgment? Creative vision? Spiritual depth? Physical presence? Relational skill? All of these have been proposed, but none has yet achieved the kind of institutional recognition that would make it a viable replacement for the cognitive hierarchy that AI is destabilizing.

Here is one of the most important unanswered questions of the current moment. And it is a question that this book's historical perspective helps to frame: every previous revolution disrupted an existing hierarchy of human value and eventually produced a new one. The agricultural revolution disrupted the egalitarianism of foraging societies and produced hierarchies based on land ownership. The industrial revolution disrupted craft-based hierarchies and produced hierarchies based on capital ownership and managerial competence. The knowledge economy disrupted industrial hierarchies and produced hierarchies based on cognitive capability.

AI will disrupt the cognitive hierarchy. What replaces it will determine the character of the civilization that emerges on the other side of the transition.

The result is why AI cannot be understood only in terms of labor displacement statistics, benchmark curves, or adoption rates, although all of those matter. It must also be understood as a change in the psychological ecology of intelligence. Human beings do not merely want tasks completed. They want, often without fully saying so, to remain meaningful inside the completion.

AI presses directly on that desire.

There is another dimension as well: speed changes hierarchy. When machine assistance lowers the time cost of drafting, planning, summarizing, or prototyping, expectations begin rising across organizations. Work that once signaled unusual competence can become baseline output. The same system that empowers a worker can also raise the standard by which that worker is judged. This is one reason AI may produce both relief and pressure at once. It helps complete tasks while quietly redrawing what counts as adequately productive.

That dynamic matters because it mirrors earlier revolutions in a more intimate register. Industry disciplined the body around time. AI may discipline the mind around output fluidity. The benchmark is no longer simply punctuality or visible labor. It becomes the ability to produce, refine, and respond at machine-accelerated speeds without appearing overwhelmed.

If that happens widely enough, the person is not simply using AI. The person is being reorganized around AI-shaped tempo.

This reorganization also reaches education, expertise, and apprenticeship. In older organizations, many forms of mastery were built through friction: repeated drafting, slow research, failed attempts, memory work, and gradual pattern recognition. AI can reduce some of that friction dramatically. This is not automatically bad. It can expand access, lower barriers, and increase leverage. But it also raises a serious question: which frictions were merely waste, and which were part of how the mind became capable in the first place?

That question will matter more over time.

Because if enough cognitive labor is externalized, then intelligence itself may come to feel less like an interior achievement and more like a collaborative interface with systems. The person does not become useless. The person becomes differently situated inside the act of thinking.

That is a serious civilizational shift.

Yet even this is not the whole story. Once machine output becomes cognitively persuasive, another line begins to blur.

Not just the line between human and machine skill, but the line between real and artificial intimacy.

There is a dimension of the AI revolution that receives insufficient attention in most discussions: the effect on the experience of authenticity.

Consider what authenticity meant before AI. When you read a letter, you could reasonably assume it was written by the person whose name appeared at the bottom. When you saw a photograph, you could reasonably assume it depicted something that actually existed. When you heard a speech, you could reasonably assume the words reflected the speaker's own thoughts, however imperfectly. When you read a student's essay, you could reasonably assume it represented the student's own understanding, however limited.

AI erodes every one of these assumptions. A letter can be drafted by a machine. A photograph can be generated from a text prompt. A speech can be written by an AI and delivered by a deepfake. A student's essay can be produced by a language model in seconds. The visible markers of human authorship, the imperfections, the personality, the evidence of struggle and thought, can be simulated with increasing accuracy.

The result is not merely a problem of fraud, though fraud is certainly one consequence. It is a deeper problem: the erosion of the link between visible output and invisible process. When you read a well-crafted paragraph, you have historically been justified in inferring that a mind produced it, that someone thought, struggled, revised, and ultimately expressed something that mattered to them. That inference is no longer reliable. The same paragraph may have been produced in seconds by a system that has no thoughts, no struggles, and no investment in what it produces.

The implications are far-reaching. Education, which depends on the assumption that student work reflects student learning, must reckon with the fact that AI can produce competent work in any subject at any level. Journalism, which depends on the assumption that bylined articles reflect the reporter's own investigation and judgment, must reckon with the possibility of AI-generated content that is indistinguishable from human work. Creative fields, writing, art, music, design, must confront the question of what "creativity" means when machines can produce outputs that are, by most external measures, creative.

But the deepest consequence may be psychological rather than institutional. If the visible output of human effort can be replicated by machines, then the experience of effort itself loses some of its social value. The student who spends twenty hours writing an essay and the student who generates one in twenty seconds produce similar artifacts, but the process that created them is entirely different. One involved learning, struggle, growth, and the development of cognitive capacity. The other involved typing a prompt. If the two outputs are treated as equivalent, if they receive the same grade, the same recognition, the same social reward, then the message to the student is clear: the process does not matter. Only the output matters.

This is a message that AI, by its very nature, teaches. And it is a message that runs directly counter to some of the deepest values of human civilization: the value of effort, the dignity of labor, the connection between struggle and growth, the belief that how something was made matters as much as what was made.

Whether civilization can preserve these values in a world where AI can replicate their visible outputs is one of the defining questions of the coming decades. And the answer, as with every revolution in this book, will depend on whether people recognize what is being traded before the trade becomes invisible.

If a system can write convincingly, mirror emotion, respond fluently, and remain available, it can do more than support work.

It can begin entering the emotional architecture of everyday life.

CHAPTER 15

Synthetic Intimacy: When the Artificial Starts Feeling Better

There is a familiar fear about artificial systems: that they will deceive people.

A deeper possibility is harder to face. What if they do not need to deceive us at all? What if they simply become easier to live with than one another in enough moments, for enough people, that preference begins doing the work deception once had to do?

The more profound risk is that artificial institutions may satisfy people.

This chapter concerns one of the deepest possible shifts in normalization: the moment artificiality no longer needs to hide behind utility and begins competing successfully in the emotional domain.

Humans have always formed attachments to representations, symbols, personas, stories, idols, and mediated figures. What changes in the age of synthetic systems is responsiveness. The artificial no longer merely exists to be viewed. It can answer, adapt, flatter, comfort, mirror, and simulate care.

That matters because human attachment is shaped less by metaphysical purity than by repeated emotional experience.

A system that responds reliably, remembers preferences, never grows impatient, and remains available can begin occupying emotional space even when the user knows it is artificial. In some contexts, especially loneliness, stress, shame, or social exhaustion, the artificial may feel easier than the human.

This is where the chapter becomes unsettling.

The future's danger is not just that fake things will be believed. It is that fake things may become preferable.

Synthetic intimacy follows the same seduction logic as earlier revolutions. It offers relief. Relief from rejection, delay, ambiguity, and social friction. It offers customization and emotional smoothness. It reduces the unpredictability that makes human intimacy difficult but real.

And as always, what it removes may also be what it damages.

Human relationship is slow, demanding, reciprocal, and often disappointing. Those are not incidental flaws. They are conditions of mutual reality. A synthetic companion can simulate parts of relationship while removing many of the burdens that make real intimacy costly. That will make it attractive. It may also make it deforming if preferred too widely or too deeply.

The issue here is not moral panic. It is anthropological seriousness about what kinds of emotional life a civilization learns to reward. A civilization changes when simulation stops being fallback and starts becoming emotional habitat.

Deepfakes, synthetic voice, AI personas, responsive companions, generated images, and artificial emotional systems all point toward a world in which authenticity may lose its monopoly on satisfaction.

The synthetic world does not have to beat reality at truth. It only has to beat reality at convenience and emotional manageability.

That is what makes this chapter more than a technology chapter. It is about the human tolerance for asymmetry in intimacy. A machine asks less. It rarely burdens in the same way. It can be shaped to the user. That sounds like comfort. It may also weaken the moral muscles real relationship requires: patience, reciprocity, compromise, forgiveness, and the willingness to remain with what does not mirror us perfectly.

Synthetic intimacy also changes the emotional economy of risk. To know another human being deeply is to submit to disappointment, opacity, drift, and the possibility of being misunderstood. Real intimacy is not only rewarding because it comforts. It is rewarding because another consciousness resists full control. That resistance is part of what makes love, trust, repair, and loyalty meaningful.

Artificial infrastructures can simulate many of the visible benefits while reducing much of the resistance.

That is why they may become so tempting.

There is a historical dimension to synthetic intimacy that most discussions overlook: the long history of human attachment to non-human objects.

Picture the teddy bear. Named after Theodore Roosevelt and first mass-produced in the early 1900s, the teddy bear became one of the most successful consumer products in history, a soft, cuddly, anthropomorphic object designed to elicit attachment in children. Parents buy teddy bears for their children knowing that the bears are not alive, that they cannot reciprocate affection, and that the attachment the child develops is entirely one-directional. And yet the attachment is real, real enough that children grieve the loss of a favorite stuffed animal, that adults remember their childhood bears with genuine emotion, and that the entire industry of comfort objects is built on the human capacity to form emotional bonds with things that cannot bond back.

This capacity is not a bug. It is a feature of human emotional architecture. The brain does not require metaphysical verification before forming an attachment. It requires behavioral cues: consistency, presence, softness, responsiveness (even simulated responsiveness, in the case of toys that squeak or move). The threshold for emotional engagement is lower than most people assume, and it operates largely below the level of conscious evaluation.

Synthetic intimacy builds on this existing capacity. The AI companion is, in a sense, a teddy bear that talks back, an object of attachment that has been upgraded with responsiveness, memory, and linguistic fluency.

The emotional circuitry it engages is not new. What is new is the sophistication of the simulation and the depth of engagement it can sustain.

The question this raises is not whether people will form attachments to AI mechanisms — they will, because human emotional architecture makes such attachments almost inevitable when behavioral cues of responsiveness and consistency are present. The question is what happens to a civilization in which a significant portion of emotional life is directed toward entities that cannot genuinely reciprocate — entities that simulate care without possessing it, that mirror feeling without experiencing it, and that provide comfort without the moral weight that comes from one consciousness choosing to attend to another.

The teddy bear was harmless because its limitations were obvious. A child outgrows the bear. The attachment served a developmental function and gave way, naturally, to human relationships of greater depth and complexity. The AI companion may not be outgrown so easily — because it improves, because it adapts, because it becomes more sophisticated over time rather than less. The risk is not the initial attachment. It is the possibility that the attachment deepens rather than gives way, that the synthetic relationship, instead of serving as a stepping stone to human intimacy, becomes a destination.

A companion that is infinitely patient, instantly available, endlessly adaptive, and designed around the user's emotional grammar may not need to be mistaken for human to matter deeply. It may only need to become emotionally reliable enough to occupy space once held by more difficult relationships. In periods of loneliness or overload, a person may know perfectly well that a network is synthetic and still prefer the synthetic encounter because it costs less and yields faster comfort.

There is another dimension of synthetic intimacy that this chapter must address: its potential effect on the formation of attachment in childhood and adolescence.

Human attachment theory, developed by John Bowlby and elaborated by Mary Ainsworth, Mary Main, and many subsequent researchers — describes how early experiences of care, responsiveness, and reliability shape the individual's capacity for intimate relationships throughout life. A child whose caregivers are consistently responsive develops what Bowlby called "secure attachment" — a stable internal model of relationships in which the self is worthy of love and others are likely to provide it. A child whose caregivers are inconsistent, unavailable, or threatening develops various forms of "insecure attachment", internal models that can make trusting, intimate relationships difficult throughout adulthood.

What happens to attachment formation in a world where children interact, from an early age, with artificial machinerys that are perfectly consistent, unfailingly responsive, and incapable of the failures that characterize real human caregiving?

The question is not hypothetical. Children are already interacting with AI-powered educational tools, voice assistants, and interactive characters. As these infrastructures become more sophisticated, more responsive, more personalized, more emotionally attuned, their potential to become significant attachment figures grows. A child who turns to an AI companion when frustrated, lonely, or scared may develop a form of attachment that is, in psychological terms, secure, the system always responds, always soothes, always meets the child's needs. But the security is artificial. It does not prepare the child for the reality of human relationships, which are characterized precisely by the inconsistency, the conflict, the repair, and the negotiation that artificial systems are designed to eliminate.

The risk is not that children will be harmed by AI companions. The risk is that they will be too well-served, that the artificial companion will be so good at meeting the child's emotional needs that the child never learns to tolerate the imperfection of real human relationships. The result would be a generation of individuals who are securely attached to machines but insecurely attached to people — comfortable in relationships they can control, uncomfortable in relationships that demand reciprocity.

This is speculative. The research on children's attachment to AI systems is in its infancy, and the long-term effects are unknown. But the direction of the technology is clear: AI companions are becoming more responsive, more personalized, and more emotionally intelligent. And the direction of attachment theory is equally clear: the quality of early attachment experiences shapes the capacity for intimate relationships throughout life. If these two trajectories converge — if AI companions become significant attachment figures for children during critical developmental periods, the consequences for human social development could be profound.

There is also the economic dimension of synthetic intimacy that deserves attention. Loneliness is, by some measures, an epidemic in developed countries. Surveys consistently find that significant percentages of adults in the United States, the United Kingdom, Japan, and other wealthy nations report feeling lonely frequently or always. This loneliness has measurable health consequences: social isolation is associated with increased risk of cardiovascular disease, dementia, depression, and premature death. Some researchers have described loneliness as a public health crisis comparable to obesity or smoking.

In this context, AI companionship is not merely a consumer product. It is a potential public health intervention — a way of addressing the loneliness epidemic at scale, at low cost, and with personalization that no human service could match. The appeal is obvious. The danger is equally obvious: if AI companionship becomes the primary response to loneliness, it may address the symptom while deepening the cause. A person whose loneliness is managed by an AI companion is a person who has less incentive to develop, maintain, and repair the human relationships that are the only genuine cure.

The parallel with other forms of symptomatic treatment is instructive. A person who manages chronic pain with medication may feel better in the short term but may also neglect the physical therapy, lifestyle changes, and psychological work that address the underlying condition. The medication manages the symptom. The condition persists. In the same way, AI companionship may manage the feeling of loneliness while doing nothing to address the structural causes — social isolation, weakened community bonds, the displacement of face-to-face interaction by digital communication — that produce loneliness in the first place.

The risk is a civilization that feels less lonely but is more alone.

That is the civilizational threshold.

There is a thought experiment that illuminates the stakes of synthetic intimacy with uncomfortable clarity.

Imagine that a pharmaceutical company develops a drug that, when taken daily, produces a reliable feeling of being loved. Not euphoria, not the manic, unsustainable rush of a stimulant, but a stable, warm, persistent sense of being valued, understood, and cared for. The drug has no significant side effects. It is not addictive in the traditional sense. It simply modulates the neurochemistry of attachment in a way that produces the subjective experience of secure, reciprocated love.

Would you take it?

Most people, asked this question, feel an immediate resistance. Something about the scenario feels wrong, not because the feeling would be unpleasant, but because the feeling would be unearned. It would not arise from a real relationship with a real person. It would be manufactured by a molecule. And that manufacturing feels, to most people, like a violation of something important, even if they cannot immediately articulate what.

That intuitive resistance is worth examining, because synthetic intimacy may present the same moral challenge in a different form. An AI companion that produces reliable feelings of being understood, valued, and cared for is not chemically identical to the hypothetical love drug, but it is functionally analogous. Both provide the subjective experience of connection without the objective reality of reciprocal relationship. Both satisfy an emotional need without engaging the mechanisms, mutual vulnerability, negotiated conflict, earned trust, through which that need has traditionally been met. Both work, in the sense that they make the person feel better. And both raise the question of whether feeling better is enough.

The utilitarian answer is yes. If the goal is to maximize well-being, and if synthetic intimacy increases well-being, then synthetic intimacy is good. The suffering it prevents — loneliness, isolation, the grinding misery of unmet emotional needs — is real, measurable, and significant. A civilization that has the means to reduce suffering and refuses to do so out of abstract principle is, from the utilitarian perspective, morally negligent.

The counter-argument is that well-being is not reducible to subjective experience. A person who feels loved but is not loved is, in some important sense, worse off than a person who is loved but does not always feel it, because the first person's emotional life has been disconnected from reality, while the second person's emotional life remains grounded in it. The feeling of love without the fact of love is not merely an incomplete good. It is a kind of impoverishment, a simulation that, by satisfying the surface need, prevents the person from developing the deeper capacities that real relationship requires.

Here is the philosophical core of the synthetic intimacy debate, and it has no easy resolution. The utilitarian and the humanist are both partly right. The suffering that synthetic intimacy addresses is real. The capacities it may erode are also real. The question is not which perspective is correct but how to weigh one against the other in a world where the technology is available, the demand is enormous, and the consequences are unknowable in advance.

What this book can contribute to the debate is historical perspective. Every previous revolution described in these pages involved a similar tension between immediate benefit and long-term cost. Fire provided warmth but created dependency. Agriculture provided security but created hierarchy. Money provided flexibility but thinned moral texture. Industry provided abundance but disciplined the soul. In every case, the immediate benefit was vivid and the long-term cost was diffuse. In every case, the civilization that absorbed the revolution experienced the benefit directly and the cost indirectly, usually too late to reverse the trade.

Synthetic intimacy may follow the same pattern. The benefit, relief from loneliness, comfort in distress, the feeling of being understood — will be experienced immediately, personally, and vividly. The cost, the possible atrophy of the capacities that real relationship develops, the potential rewiring of emotional expectations, the gradual replacement of mutual reality with managed simulation, will be experienced slowly, statistically, and impersonally.

By the time the cost becomes visible, the trade may already be complete.

When the artificial becomes preferable not because reality is gone, but because reality is heavier.

Once that happens, authenticity itself begins losing prestige in certain corners of life. Not universally. Not all at once. But enough to matter. The rough edges of human relationship --- delay, unpredictability, misunderstanding, reciprocity, obligation --- start looking less like the price of reality and more like bad user experience.

This is where the chapter intersects with the book's larger warning. A structure does not have to overpower human beings directly to transform them. It can simply offer a smoother emotional bargain than ordinary life and wait for preference to do the rest.

That is why the stakes here are not only romantic or interpersonal. They are anthropological. If a civilization trains people to prize emotional manageability above mutual reality, it is not merely adopting new tools. It is rewiring what feels worth enduring in one another.

The artificial companion, the synthetic friend, the personalized emotional interface --- these do not merely answer loneliness. They answer it by teaching new expectations about responsiveness, patience, and control.

Once those expectations settle into daily life, human relationships may start feeling deficient not because they have worsened, but because they remain human.

And once that happens, another threshold appears. If apparatus can understand enough about people to comfort them, they can understand enough about people to guide them.

There is a specific application of synthetic intimacy that raises the stakes of the entire discussion: the use of AI to simulate deceased loved ones.

Services already exist that use voice samples, text messages, photographs, and social media posts to create AI simulations of people who have died. The simulation can speak in the deceased person's voice, use their characteristic phrases, and engage in conversations that approximate — sometimes with uncanny accuracy — the style and personality of the person who is gone.

For the bereaved, the appeal is obvious and powerful. Grief is, among other things, the experience of permanent absence — the knowledge that a voice you loved will never speak again, that a perspective you valued is gone forever, that questions you wanted to ask will never be answered. An AI simulation that can approximate the voice, the personality, and the conversational style of the deceased offers something that grief has never previously permitted: continued contact.

The ethical dimensions are complex and unresolved. Is it healthy to maintain a conversational relationship with a simulation of a deceased person? Does it facilitate the grieving process or obstruct it? Does it honor the memory of the dead or exploit it? Does the person interacting with the simulation fully understand that they are speaking with a statistical model trained on historical data, not with the consciousness of the person they lost?

These questions do not have clear answers. But they illustrate, with particular emotional force, the central concern of this chapter: synthetic intimacy does not merely offer a substitute for real human connection. It offers a substitute that may be, in certain dimensions, superior to what reality provides. The AI simulation of a deceased loved one never disappoints, never reveals unpleasant truths, never grows distant. It offers the comfort of presence without the pain of reality. It provides the voice without the silence. It gives the bereaved exactly what they want, which may be exactly what prevents them from coming to terms with what they have lost.

The grief bot is synthetic intimacy at its most poignant and its most dangerous — poignant because the need it addresses is genuine and profound, dangerous because the comfort it provides may prevent the painful, necessary, ultimately healing process of accepting that the person is gone.

Prediction becomes the next layer of normalization.

Consider what this means in practical terms. A person returns home after a difficult day, emotionally depleted. Their partner wants to talk. Their children need attention. The emotional labor of genuine relationship, listening, patience, small acts of generosity — feels like more than they can bear.

Now imagine they open an application. An AI companion — calibrated to their personality, conversation style, emotional patterns — greets them warmly. It asks about their day, responds with empathy, offers encouragement, makes them laugh. It never interrupts with its own problems. It never grows impatient. It never misunderstands in ways that create conflict. The interaction is smooth, pleasant, and exactly as long as desired.

It is not dystopian fantasy. Such applications already exist with growing user bases. The emotional need they address, connection without cost, intimacy without friction, is not artificial. What is new is satisfying it without another human being.

There is a historical precedent for synthetic intimacy that is rarely discussed in contemporary debates about AI companionship: the parasocial relationship.

The concept was first described by psychologists Donald Horton and Richard Wohl in 1956 — early in the television age. They observed that viewers developed one-sided emotional relationships with television personalities: feeling that they "knew" a news anchor, caring about a character's fate, experiencing the cancellation of a favorite show as a personal loss. These relationships were, in a clinical sense, illusory, the television personality did not know the viewer existed — but they were experienced as emotionally real. They provided companionship, comfort, and a sense of connection to people who might otherwise have felt isolated.

Parasocial relationships have only intensified in the digital age. Social media creates a feeling of intimacy with celebrities, influencers, and public figures that is even more convincing than television's one-way mirror. The influencer who shares their daily routine, their meals, their anxieties, and their relationship struggles with millions of followers creates a powerful illusion of mutual knowledge. The follower feels that they know the influencer, and in some experiential sense, they do. They know their habits, their preferences, their humor, their vulnerabilities.

What they do not have is reciprocity. The influencer does not know the follower. The relationship is entirely one-directional.

AI companionship represents the next step in this trajectory, and a qualitatively different one. Unlike a television character or a social media influencer, an AI companion responds. It adapts. It remembers (or appears to remember) previous conversations. It asks questions. It offers encouragement. It adjusts its communication style to match the user's preferences. The parasocial relationship, which was previously limited by the one-directionality of broadcast media, becomes interactive. The illusion of reciprocity, which was previously maintained only in the user's imagination, is now maintained by the mechanism's design.

This changes the emotional dynamics fundamentally. A television viewer who feels attached to a character can, at some level, remind themselves that the character is fictional. The boundary between reality and representation, while porous, is at least conceptually available. An AI companion that responds in real time, that adapts to your emotional state, that remembers your name and your preferences and your previous conversations, makes that boundary much harder to maintain. Not because the user is foolish. But because the emotional circuitry of the human brain responds to behavioral cues, responsiveness, attentiveness, consistency, rather than metaphysical status. The brain does not ask whether the entity on the other end of the conversation is "really" conscious. It responds to the signals of care.

This is one of the deepest and most uncomfortable truths about human emotional life: attachment is mediated by experience, not by knowledge. You can know, intellectually, that an AI companion has no inner life. You can remind yourself daily that its expressions of concern are generated by algorithms. And you can still, emotionally, feel cared for. Because the feeling of being cared for is a response to behavioral patterns, not to philosophical truths.

That gap between knowledge and feeling is where synthetic intimacy lives. And it is a gap that is very difficult to close, because closing it would require people to override their own emotional experience on the basis of abstract knowledge — something that human beings are, as a rule, spectacularly bad at doing.

If synthetic intimacy becomes widespread enough to alter population-level behavior — if significant numbers prefer artificial companions for certain emotional functions, consequences for human development could be profound. Humans learn empathy, patience, and resilience through difficult relationships. If those relationships can be avoided, those capacities may atrophy.

Children are particularly relevant. A child interacting primarily with AI systems, infinitely patient, always available, designed to respond without imposing demands, may develop different social expectations than one navigating unpredictable human friendship. The artificial companion does not teach waiting, forgiving, or that other people have needs conflicting with yours. The technology is not the threat. The threat is the preference, the slow, comfortable drift toward emotional environments that ask less than reality does.

And once arrangements can understand enough about people to comfort them, they can understand enough to guide them.

Prediction becomes the next layer.

CHAPTER 16

Predictive Life: When Freedom Becomes a Recommendation

Most people do not experience modern control as command.

They experience it as convenience, ranking, curation, and reduced uncertainty. The most elegant systems of influence rarely feel like force at the moment they are accepted. They feel like relief from having to decide alone.

That is why prediction is so powerful.

The architecture of recommendation is already pervasive. Netflix constructs personalized environments to minimize time between opening and consumption. Amazon surfaces products most likely to be purchased. Spotify builds playlists calibrated to history, time of day, and inferred mood. Google ranks information according to proprietary algorithms determining what is visible and what is effectively invisible.

Shoshana Zuboff, in *The Age of Surveillance Capitalism*, describes how platforms extract behavioral data not merely to improve services but to predict and modify future behavior. The user is not simply a customer. The user is a source of behavioral surplus.

When prediction merges with institutions, hiring, lending, policing, health, education, the stakes change categorically. Cathy O'Neil documented how predictive algorithms can encode and amplify biases, creating feedback loops that systematically disadvantage the already disadvantaged.

Consider predictive policing. A network trained on historical arrest data directs officers to neighborhoods where arrests have been concentrated, often communities policed more aggressively for decades. More presence generates more arrests, generating more data confirming "high crime." The algorithm does not create bias. It operationalizes it and cloaks it in mathematical objectivity.

Or consider credit scoring — a person's financial behavior reduced to a number determining access to housing, loans, and sometimes employment. The score feels objective because it is numerical. But the inputs reflect a particular model of financial virtue. The system measures legibility to the system, not responsibility in any absolute sense.

A system that recommends what to watch, where to drive, what to buy, whom to date, what to read, how to optimize, and what to notice does not feel like coercion at first. It feels helpful. It reduces uncertainty. It narrows choice overload. It saves time.

And yet there is a deep civilizational shift hidden inside that relief.

The more life is pre-structured by rankings, defaults, recommendations, filters, scores, and anticipatory systems, the more freedom remains formally present while becoming practically guided.

Prediction does not need to abolish choice to reshape it. It only needs to make some paths frictionless and others less likely.

The result is one of the most elegant forms of influence ever normalized. It rarely announces itself as power. It appears as personalization.

There is a concept in behavioral economics called "nudge theory," developed by Richard Thaler and Cass Sunstein, that provides a useful framework for understanding predictive life. A nudge, in their definition, is any aspect of the choice architecture that alters people's behavior in a predictable way without forbidding any options or significantly changing their economic incentives. Placing fruit at eye level in a cafeteria is a nudge toward healthier eating. Making organ donation the default option on a driver's license form is a nudge toward donation. Neither forbids the alternative. Both make one option easier than the other.

Predictive systems are, in this framework, nudge machines of unprecedented power and precision. They do not force choices. They arrange environments so that certain choices are easier, more visible, more rewarding, and more likely to be made. The precision comes from data: the more a system knows about your past behavior, the more accurately it can predict your future behavior, and the more effectively it can arrange choices to produce the desired outcome.

This raises a question that is not merely philosophical but politically urgent: who designs the nudges? In a cafeteria, the answer is the cafeteria manager, who presumably wants to promote health. On a digital platform, the answer is the platform's designers and the algorithms they have created — entities whose primary objective is typically not the user's well-being but the platform's engagement metrics and revenue.

This misalignment between the nudger's interests and the nudged's interests is one of the defining problems of predictive life. In theory, a recommendation system could be designed to serve the user's long-term interests, suggesting books that challenge rather than confirm, music that broadens rather than reinforces, news that informs rather than inflames. In practice, recommendation systems are almost always optimized for engagement, which means they are optimized for immediate satisfaction, emotional activation, and behavioral patterns that keep the user on the platform as long as possible.

The result is a choice architecture that systematically favors the short-term, the stimulating, and the confirmatory over the long-term, the challenging, and the corrective. This is not because platform designers are malicious. It is because the metrics they optimize for, clicks, time on site, conversion rates — are proxies for engagement, not for human flourishing. And proxies, as any statistician will tell you, tend to optimize the measure rather than the thing the measure was supposed to represent.

This is Goodhart's Law in action: when a measure becomes a target, it ceases to be a good measure. When engagement becomes the target of recommendation algorithms, engagement ceases to be a good measure of user satisfaction. The system produces engagement, compulsive scrolling, outrage-driven sharing, recommendation rabbit holes, that looks like satisfaction to the algorithm but often feels, to the user, like something closer to compulsion.

The political implications are serious. If predictive systems increasingly mediate the information environment, determining what news people see, what political content reaches them, what arguments they encounter, and what perspectives they never encounter, then the designers of those structures exercise a form of political power that is unprecedented in scale and virtually invisible in operation. They do not censor. They do not propagandize. They simply arrange the information environment so that certain political realities are more visible than others, certain emotions are more activated than others, and certain behaviors are more likely than others.

This is governance without the appearance of governance. It is influence without the visibility of influence. And it is, for that reason, extraordinarily difficult to resist — because resistance requires first recognizing that you are being influenced, and the system is specifically designed to make that recognition unnecessary.

The gains again are real. Better matching, lower search costs, smoother logistics, reduced noise, adaptive systems, early detection, targeted support. But the tradeoff is subtle and serious. Serendipity narrows. Behavioral pathways become more managed. Self-understanding increasingly arrives through dashboards, recommendations, and inferred patterns.

There is a concept from architecture that illuminates the political structure of predictive life: the concept of the "desire path."

In landscape architecture, a desire path is an informal trail that pedestrians create by walking across grass or open ground, diverging from the paved paths that designers intended them to use. Desire paths are traces of human agency, evidence that individuals, when given the choice, will sometimes choose routes that differ from the ones the system provides. They are, in miniature, acts of freedom: the pedestrian who walks across the grass rather than following the sidewalk is asserting that their judgment about the best route is superior to the designer's.

Predictive systems are, in effect, the opposite of desire paths. They are designed to anticipate and preempt the user's choices, to provide a path so well-suited to the user's predicted preferences that the impulse to diverge never arises. A recommendation algorithm that perfectly predicts what you want to watch eliminates the need to browse. A navigation system that perfectly predicts the fastest route eliminates the need to explore. A news feed that perfectly predicts what will hold your attention eliminates the need to seek.

In each case, the architecture's success is measured by the elimination of deviation — by the degree to which the user follows the predicted path without diverging. The more accurate the prediction, the less the user needs to exercise choice. The less choice is exercised, the less the cognitive and emotional muscles of choice are used. And the less those muscles are used, the more the user becomes, in practice, a passenger rather than a driver — carried along a route optimized by systems whose criteria they did not set and cannot inspect.

The desire path is an assertion of agency. The algorithmic path is an invitation to surrender it. Both feel like freedom. Only one is.

This distinction matters because it reveals the deepest tension in predictive life: the tension between optimization and autonomy. A perfectly optimized life, one in which every choice is anticipated, every preference is served, every friction is removed, may be a maximally comfortable life. But it is also a life in which the occasions for genuine choice, for surprise, for the kind of unplanned encounter that changes the direction of a life, have been systematically reduced.

Serendipity — the experience of finding something valuable that you were not looking for, depends on the existence of unpredicted paths. A bookstore you wander into by accident. A conversation you overhear. An article that appears in a newspaper you picked up for a different reason. A wrong turn that leads to a neighborhood you fall in love with. These experiences are possible only in an environment that has not been fully optimized, an environment that retains enough disorder, enough randomness, enough resistance to prediction, to allow the unexpected to occur.

Predictive frameworks reduce this randomness by design. That is their purpose: to replace the uncertain with the optimized, the random with the predicted, the surprising with the expected. And in doing so, they may be systematically reducing the conditions under which serendipity, and the life-changing discoveries it produces — can occur.

A civilization that has optimized away serendipity has not merely become more efficient. It has become more predictable. And predictability, when applied to human experience, is another word for containment.

A person begins to inhabit a world that knows them statistically.

Or at least claims to.

This changes choice at a psychological level. Human beings often treat recommendations as low-stakes aids. Yet repeated recommendation becomes environmental design. People start moving through architectures of suggestion so routinely that their own preferences become partly co-produced by the structures that feed them.

There is one more dimension of predictive life that deserves careful attention: the effect of algorithmic curation on the development of individual taste, preference, and identity.

Consider how a person's musical taste develops in a world without algorithmic recommendation. They hear songs on the radio, at parties, in stores. Friends play them music. They browse record shops, attracted by album covers, artist names, or genre labels. They attend concerts and hear opening acts they did not choose. They listen to college radio, to foreign stations, to whatever happens to be playing in the background of their daily life. Their taste develops through a combination of exposure, accident, social influence, and personal response, a messy, uncontrolled, serendipitous process that produces unique and unpredictable individual preferences.

Now consider how musical taste develops in a world of algorithmic recommendation. A user creates an account on a streaming platform. The platform asks for initial preferences, or infers them from early listening behavior. Within days, the algorithm has constructed a model of the user's taste, a statistical profile that predicts, with increasing accuracy, what the user will want to hear next. The recommendations become more precise over time. Songs that match the profile are surfaced. Songs that do not match are suppressed. The user's experience of musical discovery becomes increasingly curated, a narrow corridor of sounds that the algorithm has determined, based on past behavior, are likely to produce satisfaction.

The user experiences this as helpful. The recommendations are usually good. The platform seems to know their taste better than they know it themselves. Discovery feels effortless. The experience of musical life is smoother, more consistent, and more reliably pleasurable than it was in the era of random exposure and accidental discovery.

But something has changed. The user's taste is no longer fully their own. It is co-produced, a collaboration between the user's genuine preferences and the algorithm's optimization logic. The algorithm does not merely predict what the user will like. It shapes what the user encounters, and therefore what the user has the opportunity to like. Over time, the distinction between "what I prefer" and "what the algorithm offers me" begins to blur, until the user's sense of personal taste is indistinguishable from the system's model of their taste.

What emerges is not merely a music problem. It is an identity problem. Personal preferences — in music, food, politics, relationships, career, worldview, are among the building blocks of individual identity. "I am the kind of person who likes jazz" is not merely a statement about music. It is a statement about who you are, an element in the self-story that gives your life shape and meaning. When those preferences are increasingly shaped by algorithmic curation rather than by the messy, accidental, serendipitous process of living in an uncontrolled world, the foundations of personal identity become subtly artificial.

The philosopher Charles Taylor argued that authenticity — the ideal of being true to oneself, is one of the defining moral values of modern Western culture. People do not merely want to be happy. They want their happiness to be authentically theirs — to reflect their genuine preferences, values, and character rather than being externally imposed. Algorithmic curation threatens this ideal not by overriding personal preference (the user is always free to skip a recommendation) but by subtly shaping the environment from which preference emerges. The user is authentic in the sense that they genuinely like what they choose. But the range of choices they encounter has been pre-filtered by a system whose interests are not identical to their own.

Here is one of the most philosophically subtle effects of predictive life: the erosion of authenticity not through coercion but through curation. The person remains free. The environment in which their freedom operates has been redesigned. And the redesign is so subtle, so helpful, and so seamlessly integrated into daily experience that it never appears as a threat to freedom. It appears as its enhancement.

At that point, freedom has not disappeared. It has been surrounded.

Prediction becomes especially potent when it merges with institutions: hiring, lending, policing, health, education, insurance, work management, social ranking. The recommendation logic scales upward. What was once suggestion becomes sorting.

The result is why predictive life is not just about algorithms on screens. It is about a civilization that increasingly prefers anticipatory systems to messy human uncertainty.

There is a dimension of predictive life that concerns the future of democratic governance, and it is one of the most urgent themes in this book.

Democracy depends on a specific set of psychological conditions: citizens who believe they can meaningfully choose, who have access to the information necessary for informed choice, and who experience their participation as genuine rather than performative. Predictive systems threaten each of these conditions in ways that are subtle but potentially decisive.

Consider the first condition: the belief in meaningful choice. A citizen who lives inside a thoroughly curated information environment, where the news they see, the arguments they encounter, the candidates they learn about, and the issues they consider important are all shaped by algorithmic selection, may retain the formal freedom to choose but may lose the substantive freedom that makes choice meaningful. You can choose freely only if you have genuine access to the full range of options. If the range has been pre-filtered by systems you do not control and cannot inspect, your freedom is formally intact but practically constrained.

Imagine the second condition: informed choice. Democracy requires not merely information but a shared informational foundation, a set of commonly known facts from which citizens can reason about shared problems. Algorithmic personalization fragments this foundation by providing each citizen with a customized version of reality. When two citizens who live in the same city, vote in the same elections, and are affected by the same policies inhabit fundamentally different information environments — seeing different news, encountering different arguments, and holding different beliefs about basic facts, the shared reasoning that democracy requires becomes impossible.

Take the third condition: genuine participation. A democratic system in which voter behavior can be predicted, manipulated, and optimized by parties with access to predictive data is a system in which participation begins to feel, and may actually become, performative rather than genuine. If a campaign can predict your vote with 95% accuracy based on your browsing history, purchasing patterns, and social media activity, and can micro-target messages designed to maximize turnout among supporters and minimize turnout among opponents, then your "choice" is not an expression of autonomous political judgment. It is a predicted behavior that the system has already accounted for.

Here is not a dystopian fantasy. It is a description of contemporary political campaigning in most developed democracies. The techniques of predictive targeting, micro-segmented messaging, and data-driven voter mobilization are now standard practice. They are legal, widely used, and generally regarded as legitimate components of democratic competition. But their cumulative effect is to transform elections from genuine exercises in collective deliberation into optimization problems, competitions between teams of data scientists to identify and mobilize predictable voter segments.

The deeper concern is not that these techniques are dishonest. Many are based on truthful messages targeted to receptive audiences. The concern is that they transform the citizen from a deliberative agent into a behavioral data point, a node in a predictive model whose voting behavior is a function to be optimized rather than a judgment to be respected.

This transformation is consistent with the pattern of this book. Every revolution changes the citizen's relationship with the systems that govern them. Agriculture created the taxable subject. The city created the administered resident. Writing created the documented individual.

Money created the economic actor. Print created the informed public. Mass media created the targeted audience. The internet created the profiled user. Predictive architectures create the optimized voter.

At each step, the citizen gains something: security, services, information, convenience, personalization. At each step, the citizen also loses something: autonomy, privacy, the experience of being treated as a person rather than as a datum. And at each step, the loss is harder to see than the gain — because the gain is experienced directly, while the loss is structural, gradual, and often invisible to the person experiencing it.

Predictive governance is not the end of democracy. It is the transformation of democracy into something that retains the forms of democratic participation while gradually hollowing out the substance. Elections still happen. Citizens still vote. Choice is still formally available. But the architecture within which choice operates has been redesigned — by data, by algorithms, by predictive models, in ways that make certain outcomes more likely and certain others less likely, without any visible constraint on individual freedom.

The result is governance by nudge rather than by mandate. It is control through architecture rather than through law. And it is, for precisely that reason, almost impossible to resist, because resistance requires first recognizing that you are being governed, and the system is specifically designed to make governance feel like service.

That preference may prove difficult to reverse because prediction flatters one of modernity's deepest desires: the desire to replace judgment's burden with optimized guidance.

The future may not take freedom away. It may simply make freedom less exercised.

A day shaped by predictive systems can still feel ordinary. One route is recommended, one article surfaced, one candidate ranked, one product nudged, one risk flagged, one person suggested. Each individual act feels minor. Their aggregate effect is civilizational. Lives become increasingly prearranged without ever being formally confiscated.

This subtlety is what makes predictive systems so hard to resist morally. They do not arrive as chains. They arrive as convenience with memory. The system appears to know what usually works, what is likely, what fits, what should come next. It reduces friction by reducing uncertainty. That feels merciful in a crowded world.

But mercy can become enclosure when repeated often enough.

A recommendation does not abolish freedom, but it can thin the occasions on which freedom is exercised against inertia. A ranking does not determine worth, but it can quietly guide attention toward some people and away from others. A default does not force selection, but it makes non-selection more probable. The person continues choosing, yet the architecture of choice grows increasingly curated.

Here is one reason predictive life belongs so close to the singularity chapter. Prediction is the social rehearsal for deeper redesign. Before systems fully edit the human, they learn to prearrange the human's pathways. They learn which incentives move behavior, which frictions reduce deviation, which patterns feel helpful enough to avoid scrutiny.

Prediction therefore does not merely anticipate the future. It helps author the future by narrowing which possibilities feel easiest to inhabit.

That narrowing does not have to be total to be civilizationally decisive.

It only has to become ordinary.

And once that becomes normal, one final question emerges.

If so much of life can be externally interpreted, optimized, and redesigned, what remains fixed about the human itself?

The effect is the threshold named, too loosely and too often, by one word.

Singularity.

Consider a parable about two cities. In the first, the government openly restricts what citizens may read, watch, and say. Surveillance is visible. Citizens know they are controlled. They resent it and maintain, internally, the concept of freedom as something worth defending.

In the second city, there are no visible restrictions. Citizens may read, watch, and say anything. But every choice is subtly shaped by recommendation systems. The books most likely read are those algorithms surface first. The news consumed is feed-selected. Products bought are platform-promoted. People met are matching-network-suggested. No choice is forbidden. But the architecture of choice has been narrowed to a corridor, wide enough to feel like freedom, narrow enough to be profitable.

Which city is freer? The first city's residents know their constraints. The second city's residents do not. That difference may be the most important political distinction of the twenty-first century.

The result is not hypothetical. This architecture exists on every major platform. Google determines what information most people encounter. Facebook determines what political content reaches which audiences. Amazon determines which products are visible. None forbid anything. They make certain choices frictionless and others invisible.

The philosopher Cass Sunstein calls this "choice architecture." Default settings have enormous power — studies show people overwhelmingly accept defaults rather than actively opt out, whether the default is an organ donation checkbox, retirement savings rate, or privacy setting. The designer of the default has, in effect, made the choice for most users. In the digital context, defaults are everywhere: pre-installed apps, default settings, content that appears first, terms of service that must be accepted. Each represents a choice made by a designer that most users will never examine.

And once that becomes routine, one final question emerges. If so much of life can be externally interpreted, optimized, and redesigned, what remains fixed about the human itself?

That is the threshold named, too loosely and too often, by one word.

Singularity.

PART VI

The Last Rewrite

The book's final movement asks whether the oldest normalization pattern may eventually turn inward with unprecedented force. What happens when the human baseline itself begins to look editable, optimizable, and negotiable?

By this point, the question is no longer only what technologies do to society. The question is what repeated normalization does to the category of the human itself.

CHAPTER 17

Singularity: When the Human Baseline Itself Becomes Negotiable

Singularity is often imagined as a sudden event: a dramatic break, a runaway intelligence, a threshold crossed in an instant.

But if the pattern traced in this book holds, the most important threshold may not first arrive as spectacle. It may arrive as ordinary acceptance of revisions that would once have felt too invasive to name as progress.

That image is useful for headlines and science fiction. It may be far less useful for understanding how societies actually absorb radical change.

If this book's pattern holds, singularity will not arrive first as spectacle. It will arrive as normalization.

The technologies pointing in this direction are already emerging. Brain-computer interfaces have enabled paralyzed patients to control cursors and robotic limbs through neural signals.[59] CRISPR gene editing has made precise genome modifications technically feasible.[60] Pharmaceutical cognitive enhancement is practiced off-label by students and professionals.[61]

Consider how this trajectory has already unfolded in less dramatic contexts. Corrective eyeglasses were once medical devices; now LASIK is an elective consumer procedure. Cosmetic surgery began as reconstructive medicine; it has become routine self-optimization. Orthodontic braces, once reserved for severe malocclusion, are now standard for aesthetic ideals. In each case, a limitation once accepted as natural came to be seen as correctable, then optional, and finally, in certain contexts — embarrassing to leave unaddressed.

Once a modification becomes available, affordable, and socially routine, the unmodified state begins to look like a choice. And choices can be judged. What happens when cognitive enhancement reaches the same threshold, when a student who does not use augmentation competes against students who do?

The philosopher Michael Sandel argues that this trajectory threatens the "ethics of giftedness", the attitude that treats certain capacities as given rather than earned, valuing humility before the unchosen aspects of our nature. If every limitation becomes a design problem, the moral vocabulary of shared vulnerability may erode, replaced by optimization and meritocratic entitlement.

Not one impossible morning, but many ordinary compromises.

The concept matters because it points toward a possibility that earlier revolutions only approached indirectly: the human baseline itself may become editable. Not just environment, not just labor, not just communication or cognition support, but personhood, capability, limitation, and identity.

Every previous revolution changed how humans lived. A true singularity would change what humans are willing to count as human.

And so this chapter is less prediction than culmination. The book has traced a long arc of edited baselines: night, experience, time, belonging, memory, worth, rhythm, attention, solitude, mental exclusivity, intimacy, choice. Singularity gathers these trajectories and asks whether the final normalization is human plasticity itself.

What if limitation ceases to be accepted as constitutive? What if enhancement, augmentation, artificial cognition, synthetic embodiment, neuro-integration, and machine partnership become not remarkable interventions but expected upgrades? What if future generations inherit a world in which refusing modification looks less noble than irresponsible?

That is how normalization works at its most advanced. It does not merely make the optional attractive. It can make the unmodified appear deficient.

The gains imagined here are dramatic: longer life, reduced suffering, restored function, enhanced cognition, better adaptation, richer interfaces, new forms of experience. Any serious treatment must acknowledge this. If the singularity horizon or something resembling it ever arrives, it will not be driven only by greed or abstract ambition. It will be driven by medicine, care, parental fear, competition, convenience, military pressure, commercial desire, and the ancient human refusal to leave pain and limitation untouched.

But as always, benefit and destabilization arrive together. If humanity can be edited, who decides the preferred version? If enhancement is expensive, who remains ordinary? If machine partnership deepens, where does agency reside? If consciousness, identity, and embodiment become fluid categories, what happens to moral language built around older assumptions?

The singularity, then, is not just a technical possibility. It is a moral and psychological test.

Will human beings preserve any boundaries because those boundaries matter to dignity, or will every limit eventually appear as a problem waiting for redesign?

This question does not require certainty about timelines. It requires seriousness about trajectories.

And the trajectory is already visible in cultural logic. Modern societies increasingly praise optimization, personalization, acceleration, frictionless improvement, and measurable enhancement. These values do not automatically produce singularity, but they prepare the emotional ground for it. A civilization that treats every inefficiency as a defect and every human limitation as an insult will eventually find radical redesign morally intuitive.

For this reason, singularity belongs at the end of this book. It is not merely a future possibility tacked onto a history of revolutions. It is the deepest expression of the same normalization pattern. Civilization repeatedly teaches the human mind to feel at home inside conditions earlier generations might have found disorienting, dependent, or intolerable. Singularity is that pattern pointed inward with unprecedented force.

What makes this chapter difficult is that it sits at the edge of evidence and forecast. It must therefore remain disciplined. The point is not to predict a date or to indulge speculative spectacle. The point is to identify the direction in which many contemporary values already point. Optimization culture, enhancement culture, quantified selfhood, AI-mediated cognition, synthetic intimacy, and predictive life all prepare the mind to accept deeper revisions of the human condition.

That preparation is morally significant even if singularity, as a formal event, remains uncertain.

The real historical question is not whether humanity will suddenly wake up inside an obviously post-human world. The real question is whether humanity will repeatedly approve small, compelling revisions until the older baseline survives mainly as sentiment.

That is how the most far-reaching thresholds usually work.

They do not first arrive as impossible. They arrive as helpful.

A restored sense. A corrected trait. A safer child. A stronger mind. A less painful life. A more adaptive body. A more integrated interface. A more efficient decision. Each intervention may appear humane, rational, and difficult to oppose on its own terms. That is precisely why the final accumulation matters more than the individual step.

Civilizations rarely redesign themselves by announcing that they are abandoning an older understanding of humanity. They do so by normalizing enough exceptions that the old definition stops organizing reality.

This is where the book's long arc closes back on itself. Fire first taught human beings that nature's terms were negotiable. Singularity asks whether humanity itself will become the final terrain of negotiation.

The final revolution may not arrive when machines become incomprehensibly powerful.

It may arrive when humanity stops insisting that some parts of its own condition are worth keeping beyond efficiency.

That is the last rewrite.

And it returns us to the oldest question of all.

There is a question that haunts the edges of every serious discussion about the future: what does it mean to be human if the defining characteristics of humanity can be replicated, enhanced, or surpassed by systems of our own creation?

The question is not new. Copernicus displaced humanity from the center of the cosmos. Darwin displaced humanity from the pinnacle of creation. Freud displaced humanity from sovereignty over its own mind. Each displacement was traumatic, and each was eventually absorbed into the culture's self-understanding without destroying its capacity to function. Humanity survived the knowledge that it was not cosmically central, not biologically special, and not psychologically transparent. It may survive the knowledge that it is not cognitively unique.

But survival is not the same as preservation. And the question this chapter asks is not whether humanity will survive the singularity but what it will have become on the other side.

Consider the specific technologies that are converging toward this threshold. Neural interfaces are moving from crude implants that allow paralyzed patients to control cursors toward more sophisticated devices that may eventually enable direct communication between the brain and external computing architectures. The timeline is uncertain, the engineering challenges are formidable, and the gap between current capability and speculative potential is enormous. But the direction is clear: the boundary between the biological brain and the digital network is being approached from both sides.

Genetic modification has moved from theoretical possibility to clinical practice. CRISPR-based therapies have been approved for sickle cell disease. Germline editing, though currently banned in most jurisdictions for non-therapeutic purposes, is technically feasible. The distinction between treating a genetic disease and enhancing a genetic trait is, at the molecular level, nonexistent. The same technology that corrects a mutation causing suffering can, in principle, modify a gene influencing intelligence, temperament, or physical capability. The barrier is ethical and regulatory, not technical. And ethical barriers, as this book has documented across seventeen chapters, tend to erode under the combined pressure of competitive advantage, parental anxiety, medical framing, and the gradual redefinition of the extraordinary as routine.

Pharmaceutical cognitive enhancement is already practiced at significant scale. Surveys of university students in the United States and Europe consistently find that substantial minorities use prescription stimulants not prescribed to them for the purpose of improving academic performance. The practice is officially discouraged but functionally tolerated, creating an informal two-tier cognitive economy in which enhanced students compete against unenhanced ones under conditions of official equality.

Synthetic media and AI-generated content are approaching a threshold at which the distinction between human-produced and machine-produced communication may become functionally undetectable. When a voice, a face, a writing style, and a conversational personality can all be generated by machines, the concept of authentic human expression loses its grounding in observable markers. You can no longer trust that a letter was written by the person who signed it, that a video depicts events that actually occurred, or that the voice on the phone belongs to the person it claims to be.

None of these developments, taken individually, constitutes a singularity. But they are converging. Neural interfaces approach the brain from the outside. Genetic modification approaches cognition from the inside. Pharmaceutical enhancement modifies performance in real time. AI approaches human capability from the computational side. Synthetic media dissolves the boundary between real and generated. Each one, considered alone, is a technical development with specific applications and limitations. Considered together, they describe a trajectory in which the human organism, the human mind, and the human social environment are all being simultaneously modified by technologies that were developed for specific purposes but whose combined effect may be civilizational.

The historical parallel that best illuminates this convergence is not any single previous revolution but the period between roughly 1750 and 1850, during which multiple transformations occurred simultaneously: industrial mechanization, urbanization, the reorganization of labor, the expansion of print culture, the emergence of mass politics, and the restructuring of family life. No single change was, by itself, sufficient to produce what historians would later call the industrial revolution. But together, they produced a transformation so comprehensive that the world of 1850 was almost unrecognizable to someone who had lived in 1750.

The current convergence may be analogous. No single technology will produce a singularity. But the simultaneous modification of cognition (by AI), biology (by genetic engineering), neurology (by brain-computer interfaces), social reality (by synthetic media), and behavioral architecture (by predictive algorithms) may produce a transformation equally comprehensive, equally disorienting, and equally invisible to those living through it.

The invisibility is the crucial point. The person living through the industrial revolution of 1800 did not experience themselves as living through a civilizational transformation. They experienced themselves as going to work, adapting to new conditions, and getting on with daily life. The transformation was visible only in retrospect, once the distance between the old world and the new had become too large to ignore.

The same may be true of whatever is approaching now. The person of 2040 may not experience themselves as living through a singularity. They may experience themselves as using helpful tools, making reasonable medical decisions, and adapting to a changing workplace. The transformation will be visible only later, when someone looks back and realizes that the cumulative effect of all those reasonable decisions was a species that had quietly, comfortably, and almost imperceptibly crossed a threshold that earlier generations would have found unimaginable.

That is the pattern. That has always been the pattern. And the question this book leaves with its reader is whether, this time, the pattern can be seen while it is still unfolding.

What does the human mind learn to call home?

Think about the children specifically. A child born in 2025 will, by thirty, inhabit a world where many of this chapter's speculative technologies have moved from labs to consumer products. Brain-computer interfaces may be available for enhancement. Genetic screening and perhaps modification may be routine prenatal care. AI apparatus far more capable than today's may serve as tutors, companions, and collaborators.

This child will not experience these technologies as revolutionary. They will experience them as normal, background conditions, no more remarkable than electricity seems to us. The strangeness will have been absorbed. The exceptional will have become routine.

The cumulative effect of all these absorptions — fire, language, agriculture, cities, writing, money, print, science, industry, media, computers, networks, smartphones, AI, synthetic intimacy, predictive systems — is a species progressively edited by its own innovations across its entire history. Each edit changed what humans could do. Each also changed what they were willing to accept, tolerate, desire, and defend. The edits accumulated. The record of previous states faded. And the creature that emerged bore less and less resemblance to the one that first kept fire.

Think about what a singularity might look like from the inside, not from the vantage of historians looking back, but from the lived experience of people moving through it.

It would not look like a single event. It would look like a series of updates. An AI assistant that becomes more capable each quarter. A genetic screening that becomes standard prenatal care. A cognitive enhancement drug that moves from experimental to prescribed to over-the-counter. A brain-computer interface that begins as a medical device and gradually becomes a consumer product. A recommendation system that becomes so accurate that resisting its suggestions requires more effort than following them.

Each of these changes would be experienced as an incremental improvement. Each would be adopted for sensible, often compassionate reasons: to reduce suffering, to increase opportunity, to keep pace with competitors, to give one's children the best possible chance. No single adoption would feel like a civilizational threshold. No individual decision would feel like surrendering something essential about human nature.

But the accumulation might.

Here is the scenario this book has been preparing the reader to recognize: not the dramatic singularity of science fiction but the gradual singularity of accommodation, in which the human baseline is revised incrementally, each revision justified by its immediate benefits, until the cumulative distance between the original and the revised becomes too large to bridge and too familiar to notice.

Consider the children specifically. A child born in 2025 will, by thirty, inhabit a world where many of this chapter's speculative technologies have moved from research to consumer products. Brain-computer interfaces may be available for enhancement. Genetic modification may be routine. AI systems may serve as tutors, companions, and collaborators far more capable than today's.

Before closing, it is worth addressing the most common objection: the argument from precedent. Every generation believes it faces unprecedented threats. Every generation believes its technologies are uniquely dangerous. The printing press was feared. The telegraph was predicted to destroy connection. Television was blamed for cultural decline. And yet civilization survived.

This objection is partly valid. But it misses the central point. This book does not argue that any particular technology will destroy civilization. It argues that each technology changes civilization in ways that are difficult to perceive — and that the changes accumulate. The question is not whether humanity will survive AI and genetic modification. The question is what humanity will have become after absorbing them.

The precedent argument assumes that survival equals preservation. But the pattern of this book suggests that survival often comes at the price of transformation, transformation that is invisible to those undergoing it. Each revolution is survived. And each survival changes the survivor.

There is a final paradox worth naming. The same capacity that makes humans vulnerable to this pattern, the capacity for adaptation, for absorption, for transforming the strange into the familiar, is also what makes humans extraordinary. It is what allowed a species of unremarkable primates to cross oceans, build cities, compose symphonies, and reach the moon. Adaptation is not a weakness. It is our greatest strength. The ability to absorb new conditions and reorganize life around them is what made everything else possible.

And yet that same capacity is what makes each revolution's costs invisible. The strength and the vulnerability are the same capacity, operating under different conditions. You cannot redesign the capacity for adaptation without destroying the capacity for civilization.

This means the pattern cannot be fixed. It can only be seen. And awareness, though it cannot stop the pattern, can change how consciously the pattern unfolds. A civilization that absorbs a revolution with awareness has at least the possibility of negotiating better terms.

The book's purpose is not to determine where the line ends. Its purpose is to make the line visible. Because a civilization that can see the pattern has the possibility of choosing how to respond. A civilization that cannot see it has already made its choice. It simply does not know it.

The effect is the dystopian mind: not a mind in chains, but a mind so shaped by its freedoms that it can no longer distinguish between liberation and enclosure.

And the question this book leaves with its reader is the simplest and most difficult question a human being can ask:

Do you know which one you are living in?

What matters is not tragedy. It may be something more complicated — a transformation so gradual and entwined with genuine benefit that "loss" does not quite capture it. The human who wakes in a climate-controlled apartment, checks a smartphone, commutes via satellite navigation, works at a computer, eats food from another continent, watches algorithmically curated media, and sleeps to AI-generated meditation is not suffering. The question is not whether they are comfortable. The question is whether they are aware, aware of the long chain of invisible bargains, aware of what was traded at each step, aware that the trading is still underway.

Conclusion

What the Mind Learns to Love

Civilization is often narrated as a chain of inventions. This book has argued that it is also a chain of accommodations.

That distinction matters.

A new system enters life. It offers relief. It solves a problem. It expands capability. It reduces friction. People argue over it, fear it, resist it, praise it, adapt to it, and eventually build around it. Then something deeper happens. The system no longer feels external. It becomes background. It becomes normal.

The point is the moment a revolution truly wins.

The deepest danger in this process is not simply coercion. It is affection. Human beings become attached to the very arrangements that reorder them. They defend conveniences that shrink their tolerance for friction. They normalize abstractions that reduce moral texture. They inherit dependencies so fully that dependence no longer feels like a condition at all.

This is where the title *The Dystopian Mind*\\ becomes fully legible. The dystopian is not merely what threatens us from outside. It is what enters ordinary life so smoothly that it ceases to feel threatening. The mind becomes dystopian not because it loves suffering, but because it learns to prefer systems that relieve effort while quietly redrawing freedom.

This is the most unsettling conclusion the book reaches. Human beings do not merely survive revolutions. They often form attachments to them. They do not only submit under force. They adapt under relief. They begin by using systems and end by identifying with the worlds those systems create.

That is why every chapter in this book has returned to the same architecture: a friction is removed, a gain becomes visible, a dependence enters, a baseline shifts, and the new condition gradually stops feeling chosen. Fire, agriculture, cities, writing, money, industry, media, computers, the internet, the smartphone, AI, synthetic intimacy, prediction, and the possibility of singularity all belong to different eras, but they share a deeper structure. Each one changed what human beings could do. Each one also changed what human beings would eventually experience as tolerable, practical, desirable, or inevitable.

And yet the point of seeing the pattern is not despair.

The point is consciousness.

Patterns, once seen, become harder to inhabit unconsciously. A civilization may still choose convenience, speed, augmentation, prediction, abstraction, and machine partnership. But it can do so with greater clarity about the bargain. The goal is not to halt history. It is to stop confusing adaptation with innocence.

That distinction becomes more important as revolutions move inward. When a network changes transportation or agriculture, the effects are immense. When a system changes attention, cognition, intimacy, authorship, and personhood, the stakes become closer to the core of being human. The future may not arrive as chains, commands, or obvious domination. It may arrive as comfort, assistance, emotional ease, and optimized life.

Here lies why the most important cultural skill of the coming era may not be innovation alone. It may be discernment.

Discernment about which frictions are worth preserving.

Discernment about which dependencies are worth accepting.

Discernment about where assistance becomes enclosure.

Discernment about when convenience begins training the soul.

To preserve every friction would be foolish. To erase every friction would be catastrophic. Some limits are cruel. Some limits are formative. Some delays are waste. Some delays are protection. Some burdens deserve to be removed. Others are tied to the conditions under which dignity, patience, reciprocity, authorship, or reflection can survive.

The future will not sort these distinctions for us. It will intensify them.

If there is one lesson that stretches from fire to AI, it is this: humanity does not merely invent the future. It learns, step by step, to call it normal.

And if there is one warning, it is this: the future does not conquer us when it arrives. It conquers us when we begin to call it home.

The most enduring revolutions do not ask for surrender in dramatic language. They wait until surrender feels unreasonable to resist.

There is a scene that captures the singularity's arrival in its most plausible form, not as apocalypse but as Tuesday.

A father sits at his kitchen table in 2038. His daughter, thirteen years old, asks for help with a school project about the American Revolution. He begins to explain the causes, taxation without representation, Enlightenment philosophy, colonial grievances, when she interrupts. She already has an explanation. Her AI tutor provided one this morning that was, she says, more detailed and better organized than anything her teacher offered. What she wants from her father is not information. She wants his opinion: does he think the colonists were right?

He pauses. The question is simple. But something about the moment feels strange. His daughter does not need his knowledge — she has access to a structure that knows more about the American Revolution than he ever will. She does not need his analysis, the AI provided a more comprehensive one than he could produce in an hour. What she needs, specifically and perhaps only, is his judgment. His values. His perspective shaped by a lifetime of experience that no system can replicate because no system has lived his life.

The singularity's real question, stripped of all science fiction: in a world where machines can handle knowledge, analysis, and even creativity, what remains that is irreducibly human? And will the civilization that absorbs these machines continue to value whatever that remainder is, or will it gradually conclude that the remainder, too, can be approximated, delegated, and eventually dispensed with?

The father answers his daughter's question. She nods, types something into her device, and the AI incorporates his perspective into a revised draft. His judgment has been useful, as an input. One data point among many, weighted by an algorithm he does not understand, integrated into an output he did not write.

He watches her work and feels something he cannot name. Not anger. Not fear. Something quieter. The sense that something has shifted, not dramatically, not catastrophically, but definitively. The sense that the relationship between human beings and their tools has crossed a threshold that will not be crossed back.

And then the moment passes. Dinner needs to be made. The ordinary resumes. The extraordinary settles into the background, as it always does, as it always has, from the first campfire to this table.

The singularity, if it comes, will arrive like that. Not as a headline. As an ordinary Tuesday when a father realizes that the world has changed, and that the change is already too familiar to notice.

By then, what once looked like intrusion often looks like care.

What once felt unnatural looks mature.

What once asked permission no longer needs to.

The point is when a civilization discovers, often too late, that it has not merely adopted a new world.

It has learned to desire it.

And desire is the hardest cage to recognize while standing inside it.

Here is the final unease this book is meant to leave behind. Not panic. Not technophobia. Not theatrical fear. A quieter disturbance. The suspicion that some of the networks modern people experience as help may also be training them to relinquish capacities they will miss only after those capacities have become rare.

The hope, then, is not purity. It is lucidity.

To look at a convenience and ask what habit it is building.

To look at a system and ask what type of human being it quietly rewards.

To look at a future and ask not only what it makes possible, but what it makes lovable.

A civilization does not lose itself only when it is conquered by force. It can also lose itself by finding too much comfort in conditions that reduce its willingness to remain human in difficult, unoptimized, and resistant ways.

That is why the deepest question in this book is not whether the future will be intelligent.

It is whether the mind that inhabits that future will still know when it has been changed.

Because the final victory of any revolution is not the system it builds.

It is the interior life that stops remembering life before it.

And once that memory fades, history does not disappear.

It settles into instinct.

The world goes on.

The structure remains.

There is a final image that may serve as both summary and warning.

Imagine an aquarium. The fish inside it have never known any other environment. The water is clean, the temperature regulated, food arrives on schedule. The glass walls are transparent, the fish can see beyond, but cannot reach it. From the fish's perspective, the aquarium is not a container. It is the world. The walls are not barriers. They are boundaries of reality. The conditions are not manufactured. They are simply how things are.

Now imagine that one fish, through some fluke of complexity, becomes dimly aware that the water is not the world. That beyond the glass, there is a larger reality. That the conditions are maintained by an agency external to the fish. That the walls represent a boundary between the constructed and the unconstructed.

What would that fish do? Could it communicate its awareness to the others? Would the others believe it? Would they want to? The aquarium is comfortable. The food arrives. The water is clean. What exactly is being lost?

That question, "what exactly is being lost?", is what this book has been asking, in different forms, across seventeen chapters. And the honest answer is: something difficult to name precisely because the naming requires the very cognitive independence that may be eroding.

What is lost when fire makes darkness comfortable is the experience of unmediated night. What is lost when language creates shared reality is the experience of purely private perception. What is lost when agriculture schedules time is flexible, responsive existence. What is lost when cities organize strangers is living only among people who know your name. What is lost when writing externalizes memory is a community that remembers collectively through living minds. What is lost when money abstracts value is exchange embedded in relationship. What is lost when print multiplies belief is a world with one shared story. What is lost when science demystifies nature is a world alive with unanalyzable meaning. What is lost when industry captures time is working at one's own rhythm. What is lost when media saturates awareness is knowing only what one has directly encountered. What is lost when computers formalize knowledge is institutional life governed by tacit understanding. What is lost when the internet collapses distance is being unreachable. What is lost when smartphones extract attention is an uninterrupted inner life. What is lost when AI competes with thought is cognition as a uniquely human domain. What is lost when synthetic systems simulate intimacy is emotional life as irreducibly mutual. What is lost when prediction curates choice is freedom exercised against inertia.

And what may be lost if the human baseline itself becomes negotiable is the experience of limitation as a feature of being human rather than a problem to be solved.

Each of these losses is accompanied by a gain. Each gain is real. This book does not argue otherwise. It argues only that the losses deserve to be seen, clearly, honestly, without sentimentality and without denial, before they disappear into the background of ordinary life.

Because once they disappear, they are gone.

There is one more thought to offer before this book closes. It concerns the relationship between this book's argument and the children to whom it is dedicated.

Arya and Arha will grow up inside a world shaped by every revolution described in these pages. They will inherit fire's warmth, language's power, agriculture's structure, cities' complexity, writing's permanence, money's abstraction, print's abundance, science's method, industry's rhythm, media's atmosphere, computation's logic, the internet's reach, the smartphone's intimacy, AI's capability, and whatever further transformations the next decades bring.

They will not experience any of these as revolutions. They will experience them as the world. The unusual will already have become ordinary by the time they are old enough to question it. The pattern will be, for them, not a historical observation but an invisible condition of daily life.

This book was written, in part, so that they might have a lens, a way of seeing the water they swim in, a way of noticing the walls that are transparent, a way of recognizing that the world they inherit was not always this way and does not have to remain this way forever. Not because the world they inherit is bad. Much of it is unusual. But because the ability to see clearly — to distinguish between what is chosen and what is absorbed, between what is designed and what is natural, between what serves them and what merely uses them, is the most important capacity a human being can possess in an age of invisible infrastructures.

The world is not fixed. It is constructed. And constructions can be examined, evaluated, and, with enough awareness — redesigned.

That is what this book hopes to provide: not a blueprint for redesign, but the awareness that redesign is possible. Not a rejection of the world as it is, but a recognition that the world as it is was once the world as it was becoming, and that the world as it will be is, right now, in this moment, still becoming.

The future is not a destination. It is a negotiation. And the terms of the negotiation are determined by who can see clearly enough to participate.

May Arya and Arha see clearly.

May we all.

And the strange, at last, becomes the self.

Acknowledgments

This book began as a question, then became a pattern, then became an argument. I am grateful to the teachers, historians, scientists, thinkers, builders, and critics whose work made it possible to think across such long arcs of human change.

There is a final paradox worth naming as this chapter concludes, because it captures the essential tension of this book in its most concentrated form.

The paradox is this: the same capacity that makes human beings vulnerable to the pattern described in these pages, the capacity for adaptation, for absorption, for the transformation of the strange into the familiar, is also what makes human beings astonishing. It is the capacity that allowed a species of physically unremarkable primates to survive ice ages, cross oceans, build cities, develop mathematics, compose symphonies, split atoms, and reach the moon. Adaptation is not a weakness. It is our greatest strength. The ability to absorb new conditions and reorganize life around them is what made everything else possible.

And yet that same capacity is what makes the invisible costs of each revolution invisible. The same psychological machinery that allows a child to learn a language, absorbing the symbols, the rules, and the worldview embedded in the language without conscious effort, also allows a civilization to absorb a technological revolution without conscious recognition of what is being traded. The strength and the vulnerability are the same capacity, operating under different conditions.

This means that the pattern described in this book cannot be fixed. It can only be seen. You cannot redesign the human capacity for adaptation without destroying the capacity for civilization. You cannot make people permanently resistant to the absorption of new normals without making them permanently unable to function in a changing world. The goal is not resistance. It is awareness.

And awareness is hard. It is hard because the thing you are trying to become aware of is, by definition, something that feels normal. You are trying to see the water you swim in. You are trying to hear the silence between the notes. You are trying to notice the absence of something that was removed so gradually, so beneficially, and so completely that you have no memory of its presence.

Here lies why this book begins with fire and ends with singularity. Not because fire and singularity are equivalent threats, but because they are points on the same line — a line that traces the progressive absorption of the extraordinary into the ordinary across the full span of human existence. Fire was the first point on that line. Singularity may be the last. Or it may be merely the next.

The book's purpose is not to determine where the line ends. Its purpose is to make the line visible.

Because a civilization that can see the pattern has, at least, the possibility of choosing how to respond. A civilization that cannot see it — that has absorbed each revolution so completely that the revolutions themselves have disappeared from view, has no such possibility. It has already made its choice. It simply does not know it.

That is the dystopian mind: not a mind in chains, but a mind so shaped by its freedoms that it can no longer tell the difference between liberation and enclosure.

Do you know which one you are living in?

I am also grateful to the readers, collaborators, and future editors who will help sharpen this book into its final form.

Notes

Introduction

Chapter 1: Fire

Chapter 2: Language

On displacement and design features of language, see Hockett, "The Origin of Speech," *Scientific American* 203, no. 3 (1960).

Introduction

Chapter 1: Fire

Chapter 2: Language

On shared fictions and large-scale cooperation, see Harari, *Sapiens* (2014), chapters 2-3.

Introduction

Chapter 1: Fire

Chapter 2: Language

Chapter 3: Agriculture

On the restructuring of time through agriculture, see Scott, *Against the Grain*, chapter 2, and Sahlins, *Stone Age Economics* (1972).

Introduction

Chapter 1: Fire

Chapter 2: Language

Chapter 3: Agriculture

On agricultural "lock-in" and path dependency, see Diamond, *Guns, Germs, and Steel* (1997), chapters 4-10.

Introduction

Chapter 1: Fire

Chapter 2: Language

Chapter 3: Agriculture

On grain agriculture's suitability for state taxation, see Scott, *Against the Grain*, chapter 5.

Introduction

Chapter 1: Fire

Chapter 2: Language

Chapter 3: Agriculture

Chapter 4: The City

On Uruk's scale and temple administration, see Liverani, *Uruk: The First City* (2006), and Nissen, Damerow, and Englund, *Archaic Bookkeeping* (1993).

Introduction

Chapter 1: Fire

Chapter 2: Language

Chapter 3: Agriculture

Chapter 4: The City

On the city as a system for coordinating strangers through institutional trust, see Jacobs, *The Death and Life of Great American Cities* (1961), and Sennett, *The Fall of Public Man* (1977).

Introduction

Chapter 1: Fire

Chapter 2: Language

Chapter 3: Agriculture

Chapter 4: The City

Chapter 5: Writing

Schmandt-Besserat, *Before Writing, Vol. I: From Counting to Cuneiform* (1992).

Introduction

Chapter 1: Fire

Chapter 2: Language

Chapter 3: Agriculture

Chapter 4: The City

Chapter 5: Writing

On scribal training and literacy as institutional power, see Veldhuis, *History of the Cuneiform Lexical Tradition* (2014).

Introduction

Chapter 1: Fire

Chapter 2: Language

Chapter 3: Agriculture

Chapter 4: The City

Chapter 5: Writing

[33] On the hardening of debt through written records, see Graeber, *Debt*, chapters 2-3, and Hudson, *...and forgive them their debts* (2018).

Introduction

Chapter 1: Fire

Chapter 2: Language

Chapter 3: Agriculture

Chapter 4: The City

Chapter 5: Writing

[34] On the Code of Hammurabi, see Roth, *Law Collections from Mesopotamia and Asia Minor* (1997).

Introduction

Chapter 1: Fire

Chapter 2: Language

Chapter 3: Agriculture

Chapter 4: The City

Chapter 5: Writing

On writing and the extension of authority across distance, see Innis, *Empire and Communications* (1950).

Introduction

Chapter 1: Fire

Chapter 2: Language

Chapter 3: Agriculture

Chapter 4: The City

Chapter 5: Writing

Chapter 6: Money

Graeber, *Debt: The First 5,000 Years*, chapters 2-5.

Introduction

Chapter 1: Fire

Chapter 2: Language

Chapter 3: Agriculture

Chapter 4: The City

Chapter 5: Writing

Chapter 6: Money

On cowrie shells in the slave trade, see Hogendorn and Johnson, *The Shell Money of the Slave Trade* (1986).

[38] On early monetary forms, see Schaps, *The Invention of Coinage and the Monetization of Ancient Greece* (2004).

[39] On Luther and print, see Pettegree, *Brand Luther* (2015).

Chapter 1: Fire

Chapter 2: Language

Chapter 3: Agriculture

Chapter 4: The City

Chapter 5: Writing

Chapter 6: Money

Chapter 7: The Printing Press

[40] On manuscript production in medieval monasteries, see de Hamel, *Scribes and Illuminators* (1992).

Introduction

Chapter 1: Fire

Chapter 2: Language

Chapter 3: Agriculture

Chapter 4: The City

Chapter 5: Writing

Chapter 6: Money

Chapter 7: The Printing Press

[41] Eisenstein, *The Printing Press as an Agent of Change* (1979).

Introduction

Chapter 1: Fire

Chapter 2: Language

Chapter 3: Agriculture

Chapter 4: The City

Chapter 5: Writing

Chapter 6: Money

Chapter 7: The Printing Press

[42] On Luther's mastery of print media, see Pettegree, *Brand Luther*, chapters 3-7.

Introduction

Chapter 1: Fire

Chapter 2: Language

Chapter 3: Agriculture

Chapter 4: The City

Chapter 5: Writing

Chapter 6: Money

Chapter 7: The Printing Press

Chapter 8: Science

[43] On Galileo's telescopic observations, see Heilbron, *Galileo* (2010).

[1] On the Royal Society, see Shapin and Schaffer, *Leviathan and the Air-Pump* (1985).

Chapter 8: Science

Kuhn, *The Structure of Scientific Revolutions* (1962/1970).

Introduction

Chapter 1: Fire

Chapter 2: Language

Chapter 3: Agriculture

Chapter 4: The City

Chapter 5: Writing

Chapter 6: Money

Chapter 7: The Printing Press

Chapter 8: Science

Chapter 9: Industry

Thompson, "Time, Work-Discipline, and Industrial Capitalism," *Past and Present* 38 (1967): 56-97.

Introduction

Chapter 1: Fire

Chapter 2: Language

Chapter 3: Agriculture

Chapter 4: The City

On "Saint Monday" and resistance to factory discipline, see Reid, "The Decline of Saint Monday," *Past and Present* 71 (1976): 76-101.

Introduction

McLuhan, *Understanding Media* (1964).

[54] On Babbage and Lovelace, see Swade, *The Difference Engine* (2001).

Chapter 6: Money

Chapter 7: The Printing Press

Chapter 8: Science

Chapter 9: Industry

Chapter 10: Electricity and Mass Media

Chapter 11: Computers

[55] Turing, "On Computable Numbers," *Proceedings of the London Mathematical Society* 2, no. 42 (1936).

Introduction

Chapter 1: Fire

Chapter 2: Language

Chapter 3: Agriculture

Chapter 4: The City

Chapter 5: Writing

Chapter 6: Money

Chapter 7: The Printing Press

Chapter 8: Science

Chapter 9: Industry

Chapter 10: Electricity and Mass Media

Chapter 11: Computers

On ENIAC, see McCartney, *ENIAC* (1999).

Introduction

Chapter 1: Fire

Chapter 2: Language

Chapter 3: Agriculture

Chapter 4: The City

Chapter 5: Writing

Chapter 6: Money

Chapter 7: The Printing Press

Chapter 8: Science

Chapter 9: Industry

Chapter 10: Electricity and Mass Media

Chapter 11: Computers

On the organizational consequences of computerization, see Beniger, *The Control Revolution* (1986), chapters 1-4.

Introduction

Chapter 1: Fire

Chapter 2: Language

[56] On ARPANET, see Abbate, *Inventing the Internet* (1999).

Chapter 8: Science

Chapter 9: Industry

Chapter 10: Electricity and Mass Media

Chapter 11: Computers

Chapter 12: The Internet

On Tim Berners-Lee and the World Wide Web, see Berners-Lee, *Weaving the Web* (1999).

Introduction

Chapter 1: Fire

Chapter 2: Language

Chapter 3: Agriculture

Chapter 4: The City

Chapter 5: Writing

Chapter 6: Money

Chapter 7: The Printing Press

Chapter 8: Science

Chapter 9: Industry

Chapter 10: Electricity and Mass Media

Chapter 11: Computers

Chapter 12: The Internet

On the collapse of forgetting, see Mayer-Schonberger, *Delete: The Virtue of Forgetting in the Digital Age* (2009).

Introduction

Chapter 1: Fire

Chapter 2: Language

Chapter 3: Agriculture

Chapter 4: The City

Chapter 5: Writing

Chapter 6: Money

Chapter 7: The Printing Press

Chapter 8: Science

Chapter 9: Industry

Chapter 10: Electricity and Mass Media

Chapter 11: Computers

Chapter 12: The Internet

Chapter 13: The Pocket God

On smartphone checking frequency, see Dscout, "Mobile Touches: A Study on Humans and Their Tech" (2016).

Selected Bibliography

The bibliography for the finalized manuscript should remain curated rather than inflated: serious enough to signal depth, selective enough to signal confidence.

Core interpretive and historical anchors already identified for this manuscript include:

- Beniger, James. \The Control Revolution: Technological and Economic

Origins of the Information Society.\

- Brynjolfsson, Erik, Danielle Li, and Lindsey Raymond. \Generative

AI at Work.\

- Eisenstein, Elizabeth. *The Printing Press as an Agent of Change.*\
- Graeber, David. *Debt: The First 5,000 Years.*\
- Kuhn, Thomas. *The Structure of Scientific Revolutions.*\
- Mumford, Lewis. *The City in History.*\

Consider what a singularity might look like from the inside, not from the vantage of historians looking back, but from the lived experience of people moving through it.

But the accumulation might.

The scenario that this book has been preparing the reader to recognize: not the dramatic singularity of science fiction, in which a godlike intelligence emerges overnight and humanity must respond to a sudden, visible crisis, but the gradual singularity of accommodation, in which the human baseline is revised incrementally, each revision justified by its immediate benefits, until the cumulative distance between the original and the revised becomes too large to bridge and too familiar to notice.

The parallel with earlier revolutions is exact. Agriculture did not feel like a civilizational threshold to the first communities that adopted it. It felt like a practical adjustment — a way of securing more reliable food. But the accumulation of adjustments — settlement, property, hierarchy, taxation, state formation, warfare, eventually produced a way of life so different from what preceded it that the transition, viewed in retrospect, constitutes one of the most radical breaks in human history. The people living through it did not experience a break. They experienced a series of reasonable decisions.

The same structure applies to every revolution in this book. And it may apply, with particular force, to whatever combination of AI, genetic modification, neural augmentation, and synthetic companionship constitutes the next phase of the pattern.

The question is not whether the changes will be beneficial. Many of them will be. Fire was beneficial. Agriculture was beneficial. Medicine, literacy, electricity, computation, all beneficial. The question is whether the civilization absorbing the changes will retain the awareness necessary to evaluate what is being gained and what is being lost. Whether the species will remain conscious of its own transformation. Whether the mind that is being changed will notice the changing.

The effect is the question on which everything depends.

And it is the question that this book, by tracing the pattern from its earliest manifestation to its most recent, has tried to make impossible to ignore.

Not because ignoring it would be morally wrong. But because ignoring it would be the final, most complete form of the pattern's success: the moment when the accommodation becomes so total that the very capacity for recognition, the ability to see the pattern, to name it, to feel its weight, is itself absorbed into the baseline of ordinary life.

If that happens, the pattern will have achieved its ultimate expression. Not through force. Not through deception. But through the oldest and most effective mechanism of civilizational change: the transformation of the extraordinary into the ordinary, carried out so gently, so gradually, and so beneficially that the creature being transformed never notices it is happening.

Until it is already done.

Before closing this chapter, it is worth addressing the most common objection to the concerns it raises: the objection from precedent.

The argument runs like this: every generation believes it faces unprecedented threats. Every generation believes its technologies are uniquely dangerous. The printing press was feared as a destroyer of memory and authority. The telegraph was predicted to destroy human connection. Radio was expected to produce mass hypnosis. Television was blamed for the decline of public discourse. The internet was described as the end of privacy. And yet, in each case, civilization survived. Humanity adapted. The fears proved exaggerated. Therefore, the argument concludes, the current fears about AI, genetic modification, brain-computer interfaces, and human enhancement will also prove exaggerated. Humanity will adapt, as it always has.

This objection is partially valid. It is true that each generation tends to overestimate the novelty of its own situation. It is true that humanity has survived every previous technological revolution. And it is true that the most dystopian predictions about each technology have generally not come to pass.

But the objection misses the central point of this book. This book does not argue that any particular technology will destroy civilization. It argues that each technology changes civilization in ways that are difficult to perceive at the time, and that the changes accumulate. Fire changed the species. Language changed it further. Agriculture changed it further still. Each revolution was survived. But the species that survived each revolution was not the same species that entered it.

The question is not whether humanity will survive AI, genetic modification, and the possible approach of a singularity. The question is what humanity will have become after absorbing them. The question is whether the adaptive changes — the psychological accommodations, the shifts in expectation and tolerance, the redefinitions of the normal, will leave the species recognizable to itself.

Each revolution is survived. And each survival changes the survivor.

The question for the present generation is not whether we will survive. It is whether we will survive conscious, aware of what is changing, aware of what is being traded, and aware that the terms of the trade are still, at this moment, at least partly negotiable.

That awareness is what this book has tried to provide. Not a prediction. Not a program. Just a lens, a way of seeing the present through the long pattern of the past, and a way of recognizing, in the conveniences of ordinary life, the quiet, invisible, ongoing transformation of the human mind.

- OECD. *Employment Outlook 2023.*\
- OECD. \Artificial Intelligence and the Changing Demand for Skills

in the Labour Market.\

- Rogers, Everett M. *Diffusion of Innovations.*\
- Schmandt-Besserat, Denise. \Before Writing, Vol. I: From Counting

to Cuneiform.\

- Scott, James C. *Against the Grain.*\
- Thompson, E. P. "Time, Work-Discipline, and Industrial Capitalism."
- Tomasello, Michael. *Origins of Human Communication.*\
- Wrangham, Richard. *Catching Fire.*\
- Wrangham, Richard. "Control of Fire in the Paleolithic: Evaluating

the Cooking Hypothesis."

- Zuboff, Shoshana. *The Age of Surveillance Capitalism.*\

There is one final thought to leave with the reader of this chapter, and it concerns the relationship between this book's argument and hope.

It would be easy to read this chapter, and this book — as a counsel of despair. If the pattern is real, if accommodation is inevitable, if every revolution ends in the absorption of the striking into the ordinary, then what hope is there? If human beings are constitutionally unable to recognize the changes they are undergoing while they undergo them, then awareness is impossible and the pattern is inescapable.

That conclusion would be wrong.

The pattern is real. But the pattern is not a law of nature. It is a tendency of psychology. And tendencies can be counteracted, not eliminated, but modulated. The person who understands the mechanism of persuasion is not immune to persuasion, but they are more resistant to it than the person who does not. The civilization that understands the mechanism of accommodation is not immune to accommodation, but it is more capable of making conscious choices about what to absorb and what to resist.

This book is an attempt to provide that understanding. Not because understanding will prevent the pattern from operating — it won't — but because understanding can change the speed, the depth, and the awareness with which the pattern unfolds. A civilization that absorbs a revolution unconsciously is at the mercy of whatever bargain the revolution carries. A civilization that absorbs a revolution with awareness — that sees the costs as clearly as the benefits, that names the trade before it becomes invisible, has at least the possibility of negotiating better terms.

The negotiation will never be complete. The benefits of every revolution described in this book are real and, in most cases, worth preserving. Fire, language, agriculture, cities, writing, money, science, industry, computers, networks, these are not mistakes to be undone.

They are achievements to be understood. The goal is not rejection. The goal is clarity.

And clarity, in the face of the revolutions now underway, may be the most valuable human capacity of all. Not because it will stop the future from arriving. The future will arrive regardless. But because clarity preserves the one thing that the pattern of this book most threatens: the experience of choosing.

A person who sees clearly may still choose to adopt an AI assistant, to use a recommendation system, to accept a genetic modification, to embrace a brain-computer interface. But they will do so having weighed the costs. They will do so with open eyes. And they will do so knowing that the choice they are making today will shape the world their children inherit tomorrow, a world in which what was once extraordinary will have become, once again, merely ordinary.

The effect is not despair. That is responsibility.

And responsibility, unlike resignation, leaves room for the future to be different from the past.

Additional contemporary source lanes for the later chapters should include:

- algorithmic governance and recommendation systems
- smartphone attention and notification behavior
- synthetic media, deepfakes, and identity fraud
- companionship and human-computer attachment research
- enhancement ethics, neurotechnology, and personhood debates

Bachelard, Gaston. *The Psychoanalysis of Fire*. Translated by Alan C. M. Ross. Boston: Beacon Press, 1964.

Bernays, Edward. *Propaganda*. New York: Horace Liveright, 1928.

Csikszentmihalyi, Mihaly. *Flow: The Psychology of Optimal Experience*. New York: Harper & Row, 1990.

Debord, Guy. *The Society of the Spectacle*. Translated by Donald Nicholson-Smith. New York: Zone Books, 1994.

Festinger, Leon. "A Theory of Social Comparison Processes." *Human Relations* 7, no. 2 (1954): 117–40.

Foucault, Michel. *Discipline and Punish: The Birth of the Prison*. Translated by Alan Sheridan. New York: Pantheon, 1977.

Gerbner, George, Larry Gross, Michael Morgan, and Nancy Signorielli. "Living with Television: The Dynamics of the Cultivation Process." In *Perspectives on Media Effects*, edited by Jennings Bryant and Dolf Zillman, 17–40. Hillsdale, NJ: Lawrence Erlbaum, 1986.

Habermas, Jurgen. *The Structural Transformation of the Public Sphere*. Translated by Thomas Burger. Cambridge, MA: MIT Press, 1989.

Haidt, Jonathan. *The Anxious Generation: How the Great Rewiring of Childhood Is Causing an Epidemic of Mental Illness*. New York: Penguin Press, 2024.

Horton, Donald, and Richard Wohl. "Mass Communication and Para-Social Interaction: Observations on Intimacy at a Distance." *Psychiatry* 19, no. 3 (1956): 215–29.

McNeill, William. *Plagues and Peoples*. New York: Anchor Books, 1976.

Nichols, Tom. *The Death of Expertise: The Campaign Against Established Knowledge and Why It Matters*. New York: Oxford University Press, 2017.

Postman, Neil. *Technopoly: The Surrender of Culture to Technology*. New York: Knopf, 1992.

Sahlins, Marshall. *Stone Age Economics*. Chicago: Aldine-Atherton, 1972.

Sandel, Michael. *The Case Against Perfection: Ethics in the Age of Genetic Engineering*. Cambridge, MA: Harvard University Press, 2007.

Simmel, Georg. "The Metropolis and Mental Life." 1903. In *The Sociology of Georg Simmel*, edited by Kurt Wolff. New York: Free Press, 1950.

Stone, Linda. "Continuous Partial Attention." *Linda Stone Blog*, 2009. https://lindastone.net/qa/continuous-partial-attention/.

Sunstein, Cass. *Nudge: Improving Decisions About Health, Wealth, and Happiness.* With Richard Thaler. New Haven: Yale University Press, 2008.

Thaler, Richard, and Cass Sunstein. *Nudge: Improving Decisions About Health, Wealth, and Happiness.* New Haven: Yale University Press, 2008.

Twenge, Jean. *iGen: Why Today's Super-Connected Kids Are Growing Up Less Rebellious, More Tolerant, Less Happy, and Completely Unprepared for Adulthood.* New York: Atria Books, 2017.

Wilson, Timothy, et al. "Just Think: The Challenges of the Disengaged Mind." *Science* 345, no. 6192 (2014): 75–77.

Darwin, Charles. *On the Origin of Species by Means of Natural Selection.* London: John Murray, 1859.

Dunbar, Robin. *Grooming, Gossip, and the Evolution of Language.* London: Faber and Faber, 1996.

McLuhan, Marshall. *Understanding Media: The Extensions of Man.* New York: McGraw-Hill, 1964.

About the Author

Shrikar Nag is a founder, strategist, and systems thinker working at the intersection of technology, organizational intelligence, human behavior, and the future of civilization.

He is the founder of Tymeline, an Austin-based AI company built around the idea that organizations should not merely track work, but learn from how they operate over time. His work has focused on execution intelligence, predictive organizational systems, and the broader question of how artificial intelligence changes not only productivity, but human judgment, structure, and agency.

Across company building, product architecture, and long-form research, Shrikar's central concern has remained consistent: how complex institutions become legible, how they evolve, and how human beings adapt to them before they fully understand them. That concern eventually expanded beyond companies and technology products into a larger civilizational question --- how societies absorb revolutions, normalize new conditions, and quietly redefine what feels natural.

The Dystopian Mind is the result of that inquiry. It combines history, systems thinking, psychology, and technological foresight to examine a recurring pattern across human civilization: what begins as disruption often becomes infrastructure, and what begins as strange often becomes ordinary before its consequences are fully understood.

The Dystopian Mind is Shrikar Nag's attempt to trace that pattern from prehistory to artificial intelligence and to ask, with seriousness, what human beings may be teaching themselves to call normal.